Magnetic Resonance Force Microscopy and a Single-Spin Measurement

Magnetic Resonance Force Microscopy and a Single-Spin Measurement

Gennady P. Berman
Los Alamos National Laboratory, USA

Fausto Borgonovi
Universitá Cattolica, Italy
and Istituto Nazionale di Fisica Nucleare, Italy

Vyacheslav N. Gorshkov
Institute of Physics, Ukraine
and Los Alamos National Laboratory, USA

Vladimir I. Tsifrinovich
Polytechnic University, New York, USA

World Scientific

NEW JERSEY · LONDON · SINGAPORE · BEIJING · SHANGHAI · HONG KONG · TAIPEI · CHENNAI

Published by

World Scientific Publishing Co. Pte. Ltd.

5 Toh Tuck Link, Singapore 596224

USA office: 27 Warren Street, Suite 401-402, Hackensack, NJ 07601

UK office: 57 Shelton Street, Covent Garden, London WC2H 9HE

British Library Cataloguing-in-Publication Data
A catalogue record for this book is available from the British Library.

MAGNETIC RESONANCE FORCE MICROSCOPY AND A SINGLE-SPIN
MEASUREMENT

ISBN-13 978-981-256-693-5
ISBN-10 981-256-693-7

Printed in Singapore

Preface

Magnetic resonance force microscopy (MRFM) is a rapidly evolving field, which originated in 1990s and recently achieved the first detection of a single electron spin below the surface of a non-transparent solid. Further development of MRFM techniques may have a great impact on many areas of science and technology including physics, chemistry, biology, nanotechnology, spintronics, and even medicine. Scientists, engineers, and students with a variety of backgrounds may be interested in learning about MRFM.

The objective of our book "Magnetic Resonance Force Microscopy and a Single-Spin Measurement" is to describe the basic principles as well as the advanced theory of MRFM. It is a multi-level book. Even a reader who is not familiar with quantum mechanics and magnetic resonance can understand and appreciate the basic principles of MRFM. Scientists, who work in the field of quantum physics or magnetic resonance, can obtain interesting and important information about MRFM theory and its applications.

Our book does not cover all techniques and theoretical methods used in MRFM. We describe the results, which are important for understanding the basic principles of MRFM and its applications. The main attention is paid to the oscillating cantilever-driven adiabatic reversals (OSCAR) technique in MRFM, which has been used for the experimental detection of a single electron spin. The book is written by authors who took part in the exciting development of the MRFM theory, and it is based on their own research. The book may be interesting for a wide range of readers from undergraduate students to experienced scientists, who wish to be familiar with this new promising field of science.

We are especially thankful to G. Chapline, S. A. Gurvitz, P. C. Hammel, D. Rugar, J. A. Sidles, for many useful discussions and to S. Wolf of DARPA (now at the University of Virginia) for his interest in our work and his encouragement. We are grateful to B. M. Chernobrod, G. Chapline, G. D. Doolen, H. S. Goan, S. A. Gurvitz, P. C. Hammel, D. I. Kamenev, G. V. Lopez, D. V. Pelekhov, D. Rugar and A. Sutter with whom many of the results discussed in this book were derived. We thank the Center for Nonlinear Studies at LANL for interest in and encouragement of our research, and the Laboratory Directed Research & Development (LDRD) Program at LANL for continuing support of basic scientific research, including quantum computation and quantum measurement. We also thank the Quantum Institute at LANL for encouraging scientific excellence in quantum research. Our research was supported by the Department of Energy (DOE) under contract W-7405-ENG-36 and the DOE Office of Basic Energy Sciences, by the Defense Advanced Research Projects Agency (DARPA) MOSAIC Program, by the National Security Agency (NSA) and Advanced Research and Development Activity (ARDA), and by the Army Research Office (ARO).

G. P. Berman, F. Borgonovi, V. N. Gorshkov, V. I. Tsifrinovich

Los Alamos,
February 2006

Contents

1 Introduction   1

2 Spin Dynamics — Quasiclassical Description   5

3 Spin Dynamics — Quantum Description   15

4 Mechanical Vibrations of the Cantilever   25

5 Single-Spin Detection in Magnetic Force Microscopy (MFM)   33
    5.1 Static displacement of the cantilever tip (CT) 33
    5.2 Decoherence time 38

6 Transient Process in MFM — The Exact Solution of the
Master Equation   41
    6.1 Hamiltonian and master equation for the spin-CT system . . . 41
    6.2 Solution for spin diagonal matrix elements 47
    6.3 Solution for spin off-diagonal matrix elements 54

7 Periodic Spin Reversals in Magnetic Resonance Force
Microscopy (MRFM) Driven by π-Pulses   59

8 Oscillating Adiabatic Spin Reversals Driven by the
Frequency Modulated rf Field   65
    8.1 Schrödinger dynamics of the CT-spin system 66
    8.2 Decoherence and thermal diffusion for the CT 79

9 Oscillating Cantilever-Driven Adiabatic Reversals (OSCAR) Technique in MRFM   85

9.1 CT-spin dynamics: discussion and estimates 87

9.2 Experimental detection of a single spin 94

10 CT-Spin Dynamics in the OSCAR Technique   97

10.1 Quasiclassical theory: simple geometry 97

10.2 Quantum theory of the OSCAR MRFM 106

10.3 OSCAR frequency shift for a realistic setup 114

11 Magnetic Noise and Spin Relaxation in OSCAR MRFM   127

11.1 OSCAR relaxation in a spin ensemble 128

11.2 Reduction of magnetic noise 145

11.3 Simple model for quantum jumps 153

11.4 Reduction of the frequency shift due to the CT-spin
entanglement 158

12 MRFM Applications: Measurement of an Entangled State and Quantum Computation   163

12.1 MRFM measurement of an entangled spin state 163

12.2 MRFM based spin quantum computer 169

13 MRFM Techniques and Spin Diffusion   183

13.1 Spin diffusion in the presence of a nonuniform magnetic field . 184

13.2 Suppression of the spin diffusion in a spin quantum computer   193

14 Conclusion   207

14.1 Abbreviations 209

14.2 Prefixes 210

14.3 Notations 211

Bibliography 217

Index 223

Chapter 1

Introduction

Recent remarkable progress in magnetic resonance imaging (MRI) naturally raises the question about the ultimate possibility for atomic scale MRI resolution (single spin detection). However, the MRI techniques are based on the phenomenon of electromagnetic induction, which implies the detection of a macroscopic number of spins. The minimum number of nuclear spins detected by MRI techniques is about 10^{12} [1], and the minimum number of electron spins detected using electron spin resonance (ESR) techniques is about 10^7 [2]. To resolve the problem of a single spin detection Sidles [3, 4] suggested using the force detection techniques like those used in atomic force microscopy (AFM). However, the magnetic force is much smaller than the electric force detected in AFM. In order to overcome this difference Sidles proposed a combination of magnetic resonance techniques for a single spin with the mechanical resonance of an ultrasensitive cantilever.

According to this idea, a ferromagnetic particle attached to the cantilever tip (CT) will experience a magnetic force produced by a single spin. If the frequency of the spin oscillations matches the resonant frequency of the cantilever vibrations, the spin force will amplify the cantilever vibrations, which can be detected, for example, by optical methods. This method of a single spin detection was labeled magnetic resonance force microscopy (MRFM).

The idea of MRFM quickly attracted the attention of experimentalists.

Soon, the MRFM techniques were implemented by Rugar *et al.* in ESR [5] and nuclear magnetic resonance (NMR) [6], and by Zhang *et al.* in ferromagnetic resonance (FMR) [7]. Finally, in 2004, 13 years after the proposal of MRFM, Rugar *et al.* [8] announced the first detection of a single electron spin below the surface of a non-transparent material (vitreous silica), using a modified MRFM technique.

The authors of this book believe that MRFM will find many important applications in physics, chemistry, biology and medicine. Consequently we decided to write this book to explain the basic ideas of MRFM and some theoretical approaches used to describe the MRFM techniques for readers with a variety of backgrounds. We were fortunate to take part in the development of the MRFM theory, and our book is based mainly on research in which we directly participated.

The book is organized as follows. In Chapters 2–4, we give the basic information about the classical and quantum description of magnetic resonance and quantum theory of a simple harmonic motion. These chapters are written for a reader who is not familiar with magnetic resonance or coherent states in quantum mechanics but who wants to understand the theoretical approaches used in MRFM. In Chapters 5 and 6, we consider the detection of a single spin using magnetic force microscopy (MFM) without magnetic resonance. The experimental implementation of a single spin MFM is beyond current experimental capability. However, from the theoretical point of view MFM is much simpler than MRFM. In particular MFM allows us to obtain the exact analytical solution for the master equation which is impossible for MRFM. Thus, the theory of a single-spin MFM allows one to understand the spin-cantilever system. In Chapter 7, we describe one of the simplest MRFM techniques which could be used for a single-spin measurement. In this technique a periodic sequence of the *rf* π-pulses drives periodic spin reversals, which, in turn, drive the cantilever vibrations. In Chapter 8, we describe a more sophisticated technique, in which cyclic adiabatic spin reversals are driven by a frequency modulated *rf* field. This technique has been widely used in MRFM experiments with macroscopic ensembles of electron and nuclear spins. Chapters 9, 10 and 11 are devoted to the oscillating cantilever driven adiabatic reversals (OSCAR) technique, which was used in [8] for single-spin detection. In this technique spin cyclic adiabatic reversals are driven by the cantilever vibrations in the presence of an *rf* field. In turn, spin reversals cause a frequency shift of the cantilever vibrations, which can be detected with high precision. In Chapter 11, Section 4, we propose a new experiment for measuring the characteristic time-scale for the collapse of the spin-cantilever wave function. In Chapter 12, we discuss possible applications of MRFM to the measurement of spin entangled states and to quantum computation. In Chapter 13, we consider the application of highly nonuniform magnetic fields used in MRFM techniques for the suppression of the spin diffusion and relaxation.

Our book has a multi-level structure. Even a reader who is not familiar with magnetic resonance and quantum mechanics can understand the basic principles of the MRFM if he or she will read Chapter 2 (the quasiclassical theory of the magnetic resonance) and skip all "quantum sections" of the book. The next level includes the readers who are familiar with magnetic resonance but not familiar with quantum mechanics. They may skip Chapter 2 and all "quantum sections" of the book. The readers who are not familiar with the master equation may skip the corresponding sections but still understand the quantum theory of MRFM.

We would like to mention that in addition to MRFM there are other approaches to the single-spin measurement in condensed matter. One of them relies on the optical detection of magnetic resonance. (See, for example, the review by Köhler [9] and the recent paper of Jelezko *et al.* [10].) As an example in the fluorescence-detected magnetic resonance (FDMR) technique a single molecule is excited with a laser to a metastable paramagnetic state. The magnetic resonance in the metastable state under the action of the *rf* field is detected via a change in the fluorescence intensity. Another approach utilizes the scanned tunneling microscopy (STM). As an example, Manassen *et al.* [11] reported modulation of the tunneling current with the Larmor frequency of the localized spin of an individual iron atom in silicon in the presence of a small permanent magnetic field. Recently, Elzerman *et al.* [12] demonstrated the electrical measurement of the spin state of an individual electron spin in a semiconductor quantum dot. They used spin-to-charge conversion of a single electron in a dot, and detected a single-electron charge using a quantum point contact. Xiao *et al.* [13] reported the electrical detection of magnetic resonance under the action of an *rf* field for a structural single electron paramagnetic defect near the Si/SiO_2 interface. They also used spin-to-charge conversion. The electric charge was measured using a silicon field-effect transistor.

We do not intend here to give the full list of articles. All of the single-spin measurement approaches may find (or may already have found) important applications in science and technology. However, so far, MRFM is the only approach which has the potential to detect a single spin and to measure the state of the spin localized below the surface of a non-transparent material.

Chapter 2

Spin Dynamics — Quasiclassical Description

While a spin is a quantum object, in many cases its dynamics can be successfully described using quasiclassical theory. The main property of an electron or nuclear spin is the following: the spin's magnetic moment $\vec{\mu}$ is parallel to the spin $\vec{S}$. We can write $\vec{\mu} = \pm\gamma\vec{S}$, where γ is the magnitude of the gyromagnetic ratio. The positive sign in this equation corresponds to the proton's spin and many other nuclear spins. The negative sign corresponds to the electron's spin and also some nuclear spins. We will consider an electron spin with a negative gyromagnetic ratio. The direction of the electron magnetic moment is opposite to the direction of the spin.

It is well known that a uniform magnetic field $\vec{B}$ does not produce a net force on a magnetic moment: the force acting on the positive North pole is balanced by the force acting on the negative South pole. The torque $\vec{\tau}$ produced by the magnetic field is given by:

$$\vec{\tau} = \vec{\mu} \times \vec{B}. \qquad (2.1)$$

The rate of change of the spin direction is equal to the torque:

$$\dot{\vec{S}} = \vec{\tau} = \vec{\mu} \times \vec{B}. \qquad (2.2)$$

Now, multiplying both sides of this equation by $(-\gamma)$, we derive the quasi-classical equation of motion for the magnetic moment:

$$\dot{\vec{\mu}} = -\gamma \vec{\mu} \times \vec{B}. \qquad (2.3)$$

We will write this equation in terms of its Cartesian components:

$$
\begin{aligned}
\dot{\mu}_x &= -\gamma \left(\mu_y B_z - \mu_z B_y \right), \\
\dot{\mu}_y &= -\gamma \left(\mu_z B_x - \mu_x B_z \right), \\
\dot{\mu}_z &= -\gamma \left(\mu_x B_y - \mu_y B_x \right).
\end{aligned}
\qquad (2.4)
$$

Let $\vec{B} = \vec{B}_{ext}$, where $\vec{B}_{ext}$ is a permanent external magnetic field, which points in the positive z-direction. Then Eqs. (2.4) can be rewritten as:

$$
\begin{aligned}
\dot{\mu}_x &= -\gamma \mu_y B_{ext}, \\
\dot{\mu}_y &= \gamma \mu_x B_{ext}, \\
\dot{\mu}_z &= 0.
\end{aligned}
\qquad (2.5)
$$

There are two equilibrium directions for the vector $\vec{\mu}$: the positive z-direction and the negative z-direction. The first case $\mu_z = \mu$ corresponds to the minimum magnetic energy

$$U_m = -\vec{B}_{ext} \cdot \vec{\mu} = -B_{ext}\, \mu. \qquad (2.6)$$

Consider the case when the transversal component of $\vec{\mu}$ is not equal to zero:

$$\mu_\perp = \left(\mu_x^2 + \mu_y^2 \right)^{1/2} \neq 0. \qquad (2.7)$$

We will multiply the second equation in (2.5) by i and add the left and the right sides of the first and second equations. Using the notation:

$$\mu_\pm = \mu_x \pm i\mu_y, \qquad (2.8)$$

we obtain the equation

$$\dot{\mu}_+ = i\gamma B_{ext}\, \mu_+. \qquad (2.9)$$

The solution is

$$\mu_+(t) = \mu_+(0)\exp(i\gamma B_{ext}t). \tag{2.10}$$

Note, that the real and imaginary parts of μ_+ are the x- and y-components of the vector $\vec{\mu}$. The solution (2.10) describes the counterclockwise precession of the magnetic moment $\vec{\mu}$ about the magnetic field. This is the well known Larmor precession of the magnetic moment.

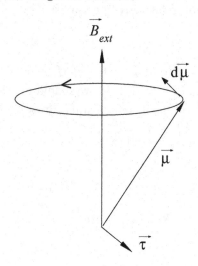

Figure 2.1: Larmor precession of the magnetic moment $\vec{\mu}$ about the magnetic field $\vec{B}_{ext}$.

The reason for the Larmor precession can be understood by looking at Fig. 2.1. When the vectors $\vec{B}_{ext}$ and $\vec{\mu}$ are in the plane of the paper, the torque $\vec{\tau} = \vec{\mu} \times \vec{B}_{ext}$ points out of the paper. The vector $d\vec{\mu} = -\gamma\tau\,dt$ has a direction opposite to the direction of the torque and points into the paper along the tangent to the circle shown in Fig. 2.1.

The z-component of the magnetic moment, i.e. the component of $\vec{\mu}$ along the direction of the magnetic field $\vec{B}_{ext}$ is an integral of motion. The other integral of motion is the magnitude of the magnetic moment:

$$\mu = \left(\mu_x^2 + \mu_y^2 + \mu_z^2\right)^{1/2}. \tag{2.11}$$

It is easy to check this by taking the derivative of μ^2:

$$\frac{d}{dt}\mu^2 = 2\vec{\mu} \cdot \dot{\vec{\mu}} = 2\vec{\mu} \cdot (-\gamma\vec{\mu} \times \vec{B}_{ext}) = 0. \qquad (2.12)$$

Next, we consider the motion of the magnetic moment in the presence of a radio frequency (rf) field $\vec{B}_1$.

Let $\vec{B}_1$ be a circularly polarized field in the $x - y$ plane rotating in the counterclockwise direction relative to the z-axis:

$$B_{1x} = B_1 \cos\omega t, \qquad B_{1y} = B_1 \sin\omega t. \qquad (2.13)$$

This expression implies that at $t = 0$ the rf field points in the positive x-direction. Now, in Eqs. (2.3) and (2.4), we should put $\vec{B} = (\vec{B}_{ext} + \vec{B}_1)$. We will rewrite Eqs. (2.4) in terms of complex quantities $\mu_\pm$ and $B_\pm = B_x \pm iB_y$:

$$\dot{\mu}_+ = i\gamma(B_z\mu_+ - B_+\mu_z),$$

$$\dot{\mu}_z = \frac{i\gamma}{2}(B_+\mu_- - B_-\mu_+). \qquad (2.14)$$

Using Eq. (2.13) and $B_z = B_{ext}$ we rewrite (2.14) in the form:

$$\dot{\mu}_+ = i\gamma(B_{ext}\mu_+ - B_1\mu_z e^{i\omega t}),$$

$$\dot{\mu}_z = \frac{i\gamma}{2}(B_1\mu_- e^{i\omega t} - B_1\mu_+ e^{-i\omega t}). \qquad (2.15)$$

To simplify these equations we should write them in the rotating system of coordinates (RSC). Rotating the coordinate system by the angle φ about the z-axis, we obtain for any vector $\vec{A}$:

$$A_x = A'_x \cos\varphi - A'_y \sin\varphi,$$
$$A_y = A'_y \cos\varphi + A'_x \sin\varphi, \qquad (2.16)$$

where the "prime" refers to RSC. We take $\varphi = \omega t$ which means that the axis x' points in the direction of the rf field. In terms of the complex variables $A_\pm$ we obtain:

$$A_\pm = A'_\pm e^{\pm i\omega t}. \qquad (2.17)$$

Thus, in the RSC Eqs. (2.15) take the form:

$$\dot{\mu}'_+ = i(\gamma B_{ext} - \omega)\mu'_+ - i\gamma B_1\mu_z,$$

$$\dot{\mu}_z = \frac{i}{2}\gamma B_1(\mu'_- - \mu'_+). \tag{2.18}$$

These equations describe the motion of the magnetic moment in the effective field $\vec{B}_{eff}$:

$$\vec{B}_{eff} = \left\{B_1, \ 0, \ B_{ext} - \frac{\omega}{\gamma}\right\}. \tag{2.19}$$

Note that the effective field in the RSC is a fixed field, which lies in the $x' - z$ plane. (See Fig. 2.2.) Thus, Eqs. (2.18) describe the Larmor precession of the

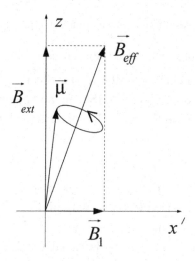

Figure 2.2: Magnetic moment in the RSC precesses about the effective magnetic field.

vector $\vec{\mu}$ about the effective field. In the "laboratory" system of coordinates (LSC) we have a complicated motion of $\vec{\mu}$: the precession about the effective field, which itself rotates about the z-axis with frequency ω. The frequency, ω_{eff}, of the precession in the RSC is, obviously :

$$\omega_{eff} = \gamma \left[B_1^2 + \left(B_{ext} - \frac{\omega}{\gamma}\right)^2\right]^{1/2}. \tag{2.20}$$

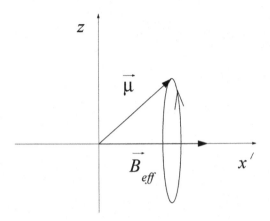

Figure 2.3: Resonance case. The magnetic moment $\vec{\mu}$ precesses about the x'-axis.

Below we consider some important special cases.

1. The magnetic moment is parallel to the effective field. There are two equilibrium directions of $\vec{\mu}$ in the RSC: in- and opposite to- the direction of $\vec{B}_{eff}$. In this case, in the LSC, the vector $\vec{\mu}$ precesses about the z-axis with frequency ω.

2. Resonance case: $\omega = \omega_e = \gamma B_{ext}$, where $\omega_e = \gamma B_{ext}$ is the electron spin resonance (ESR) frequency. In this case the effective field points in the positive x-direction (see Fig. 2.3), and $B_{eff} = B_1$. The magnetic moment precesses about the x'-axis with the frequency $\omega_{eff} = \omega_R = \gamma B_1$. The frequency $\omega_R = \gamma B_1$ is called the Rabi frequency. If the magnetic moment is perpendicular to the x'-axis it executes periodic reversals that are called Rabi oscillations. Note, that in any case the component of the vector $\vec{\mu}$ along the effective field is an integral of motion in both the RSC and the LSC.

Next, we consider the case for which the permanent magnetic field $B_{ext} = 0$, and the magnetic moment experiences the rotating magnetic field, $\vec{B}$:

$$\vec{B} = \{B \sin \Omega t, \ 0, \ B \cos \Omega t\}. \tag{2.21}$$

(See Fig. 2.4.)

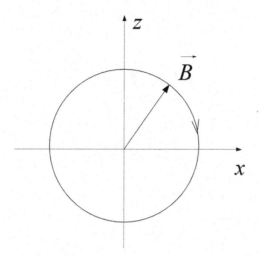

Figure 2.4: The magnetic field $\vec{B}$ rotates in the $x - z$ plane with frequency Ω.

To describe the motion of the magnetic moment, we should transfer to the system of coordinates (x', y, z') rotating with the magnetic field, $\vec{B}$. Assume z' points in the direction of $\vec{B}$. In this RSC, the effective magnetic field $\vec{B}_{eff}$ has the following components:

$$\vec{B}_{eff} = \left\{ 0, \ \frac{\Omega}{\gamma}, \ B \right\}. \tag{2.22}$$

The equations of motion in the RCS describe the precession about the effective field. (See Fig. 2.5.) The frequency of precession is:

$$\omega_{eff} = [\Omega^2 + (\gamma B)^2]^{1/2}. \tag{2.23}$$

Note that the magnetic moment periodically returns to the direction of the magnetic field $\vec{B}$.

We now consider two extreme cases for the rotating magnetic field.

1. The frequency Ω is much greater than γB (fast rotation of the magnetic field). In this case the effective magnetic field points approximately in the y-direction in Fig. 2.5. The vector $\vec{\mu}$ in the RSC precesses in the

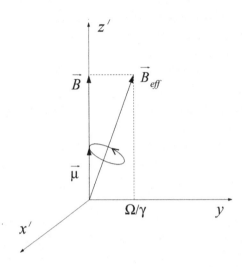

Figure 2.5: Precession of the magnetic moment $\vec{\mu}$ in the case of rotating magnetic field shown in Fig. 2.4. We assume that at $t = 0$ the vector $\vec{\mu}$ points in the positive z-direction.

plane $x' - z'$ with frequency $\omega_{eff} \approx \Omega$. In the LSC the magnetic moment does not change its direction. It points in the positive z-direction while the magnetic field rotates in the $x - z$ plane with frequency Ω. One can conclude that for $\Omega \gg \gamma B$, the magnetic moment does not "feel" the rapidly rotating magnetic field, which "averages to zero".

2. The frequency Ω is small compared to γB (adiabatic rotation of the magnetic field). In this case the effective field in the RSC (see Fig. 2.5) points approximately in the positive z'-direction. Thus, the vector $\vec{\mu}$ does not change its direction. In the LSC the magnetic moment rotates with the magnetic field $\vec{B}$ and executes periodic "adiabatic reversals". Periodic adiabatic reversals play an important role in MRFM. To implement adiabatic reversals, one does not have to rotate the net magnetic field. One of the simplest implementations is the slow modulation of the external magnetic field in the presence of the resonant rf field. (See Fig. 2.6.)

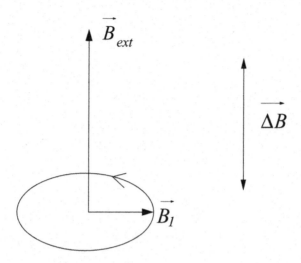

Figure 2.6: Modulation of the external field $\Delta\vec{B}$, which provides the cyclic adiabatic reversals.

When the RSC is rotating with the rf field $\vec{B}_1$, the effective magnetic field is time dependent:

$$\vec{B}_{eff} = \left\{ B_1, \ 0, \ B_{ext} - \frac{\omega}{\gamma} + \Delta B \cos \Omega t \right\}. \tag{2.24}$$

Under the resonance condition $\omega = \gamma B_{ext}$, the effective field exhibits periodic oscillations in the $x' - z$ plane (the x'-axis points in the direction of the rf field). Fig. 2.7 shows the effective magnetic field for the case when $\Delta B \gg B_1$.

Let us now assume the initial magnetic moment points in the positive z-direction. If $\Delta B \gg B_1$, the direction of the vector $\vec{\mu}$ is close to the direction of $\vec{B}_{eff}$. If $\vec{B}_{eff}$ adiabatically changes its direction, as shown in Fig. 2.7, then the magnetic moment follows $\vec{B}_{eff}$, implementing cyclic adiabatic reversals. The condition for the adiabatic motion can be derived from the following simple considerations:

The rate of change of the effective field is given by:

$$\frac{d\vec{B}_{eff}}{dt} = \{0, \ 0, \ -\Omega\Delta B \sin \Omega t\}. \tag{2.25}$$

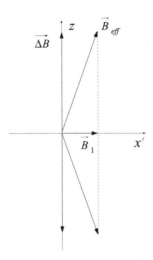

Figure 2.7: Oscillations of the effective field in the $x' - z$ plane.

Clearly, this vector has the maximum magnitude when $\Omega t = \pi(n + 1/2)$, $n = 0, 1, 2, \ldots$, i.e. when the polar angle of the effective field $\theta_{eff} = \pi/2$. The frequency of the precession in the RSC, $\omega_{eff} = \gamma B_{eff}$, has minimum at the same polar angle $\theta_{eff} = \pi/2$. If the angular speed of the effective field near the polar angle $\theta_{eff} = \pi/2$ is small compared to the precession frequency, then the condition of adiabatic motion is satisfied for any angle θ_{eff}. For $\theta_{eff} = \pi/2$ the angular displacement of the effective field is $d\theta_{eff} = |d\vec{B}_{eff}|/B_1$. The condition for the adiabatic reversals can be formulated as:

$$\frac{d\theta_{eff}}{dt} \ll \gamma B_1, \qquad (2.26)$$

or

$$\left|\frac{d\vec{B}_{eff}}{dt}\right| \ll \gamma B_1^2. \qquad (2.27)$$

Note that, instead of the modulation of the external field, one can modulate the *rf* frequency ω (see Eq. (2.24)).

Chapter 3

Spin Dynamics — Quantum Description

The main distinction between the quantum and quasiclassical descriptions of the spin is the following: in quantum mechanics the component of the spin $\vec{S}$ along any axis may take only two values $\pm 1/2$ (in units of the Planck's constant $\hbar$). In the S_z-representation the operator corresponding to the z-component of the spin is a diagonal matrix with the matrix elements $\pm 1/2$:

$$\hat{S}_z = \frac{1}{2} \begin{pmatrix} 1 & 0 \\ 0 & -1 \end{pmatrix}. \tag{3.1}$$

The operators, which correspond to the x- and y-components of the spin are the non-diagonal matrices,

$$\hat{S}_x = \frac{1}{2} \begin{pmatrix} 0 & 1 \\ 1 & 0 \end{pmatrix}, \qquad \hat{S}_y = \frac{i}{2} \begin{pmatrix} 0 & -1 \\ 1 & 0 \end{pmatrix}. \tag{3.2}$$

The operator corresponding to the spin component along the unit vector $\vec{n}$ is $\hat{\vec{S}} \cdot \vec{n}$:

$$\hat{\vec{S}} \cdot \vec{n} = \frac{1}{2} \begin{pmatrix} n_z & n_- \\ n_+ & -n_z \end{pmatrix}, \tag{3.3}$$

where $n_\pm = n_x \pm i n_y$. Using the relation, $n_+ n_- = n_x^2 + n_y^2$, one can easily prove that the eigenvalues of the operator (3.3) are $\pm 1/2$.

The wave function χ of the spin is a spinor:

$$\chi = \begin{pmatrix} c_1 \\ c_2 \end{pmatrix} = c_1 \alpha + c_2 \beta, \tag{3.4}$$

where α and β are the eigenfunctions of the operator $\hat{S}_z$ corresponding to the two values of S_z:

$$\alpha = \begin{pmatrix} 1 \\ 0 \end{pmatrix}, \qquad \beta = \begin{pmatrix} 0 \\ 1 \end{pmatrix}. \tag{3.5}$$

Solving the equation:

$$\frac{1}{2} \begin{pmatrix} n_z & n_- \\ n_+ & -n_z \end{pmatrix} \begin{pmatrix} c_1 \\ c_2 \end{pmatrix} = \pm \frac{1}{2} \begin{pmatrix} c_1 \\ c_2 \end{pmatrix} \tag{3.6}$$

one can find the normalized eigenfunctions of the operator $\hat{\vec{S}} \cdot \vec{n}$, which can be written as

$$\chi_{\frac{1}{2}} = \frac{1}{\sqrt{2}} \left(\sqrt{1 + n_z} \, \alpha + \frac{n_+}{\sqrt{1 + n_z}} \, \beta \right),$$

$$\chi_{-\frac{1}{2}} = \frac{1}{\sqrt{2}} \left(\sqrt{1 - n_z} \, \alpha - \frac{n_+}{\sqrt{1 - n_z}} \, \beta \right). \tag{3.7}$$

The subscript $\pm 1/2$ refers to the eigenvalues of the operator $\hat{\vec{S}} \cdot \vec{n}$. It is easy to verify that the eigenfunctions $\chi_{\pm 1/2}$ are orthogonal to each other: $\chi_{1/2}^\dagger \chi_{-1/2} = 0$, where "$\dagger$" means Hermitian conjugate:

$$\begin{pmatrix} c_1 \\ c_2 \end{pmatrix}^\dagger = (c_1^*, \ c_2^*). \tag{3.8}$$

Putting $n_x = 1, n_y = n_z = 0$, we will get the eigenfunction for the operator $\hat{S}_x$:

$$\chi_{\frac{1}{2}} = \frac{1}{\sqrt{2}}(\alpha + \beta), \qquad \chi_{-\frac{1}{2}} = \frac{1}{\sqrt{2}}(\alpha - \beta). \qquad (3.9)$$

In a similar way we can get the eigenfunctions for the operator $\hat{S}_y$:

$$\chi_{\frac{1}{2}} = \frac{1}{\sqrt{2}}(\alpha + i\beta), \qquad \chi_{-\frac{1}{2}} = \frac{1}{\sqrt{2}}(\alpha - i\beta). \qquad (3.10)$$

According to the measurement postulate of quantum mechanics, if an experimentalist measures the component of the spin along the unit vector $\vec{n}$ he will always transform the spin wave function to the function $\chi_{\frac{1}{2}}$ or $\chi_{-\frac{1}{2}}$, which corresponds to the two possible values of the spin component.

The spin operators $\hat{S}_x$, $\hat{S}_y$ and $\hat{S}_z$ do not commute:

$$[\hat{S}_x, \hat{S}_y] = i\hat{S}_z, \quad [\hat{S}_y, \hat{S}_z] = i\hat{S}_x, \quad [\hat{S}_z, \hat{S}_x] = i\hat{S}_y. \qquad (3.11)$$

It means that the components S_x, S_y and S_z cannot have definite values simultaneously. In other words, if an experimentalist has measured the value S_z he cannot predict what will be the result of the measurement of the component S_x or S_y.

In the S_z-representation the coefficient c_1 in (3.4) is the probability amplitude for $S_z = 1/2$ and c_2 is the probability amplitude for $S_z = -1/2$. The corresponding probabilities are given by the square of modulus $|c_1|^2$ and $|c_2|^2$. As an example, if the component of spin along the unit vector $\vec{n}$ has a definite value $S_n = 1/2$, then the wave function of the spin is given by the first equation in (3.7). If an experimentalist will measure the S_z-component for this state, he will obtain the value $S_z = 1/2$ with probability $(1 + n_z)/2$, or the value $S_z = -1/2$ with probability $(n_x^2 + n_y^2)/2(1 + n_z)$. The average value $\langle S_z \rangle$ of the spin z-component can be found as:

$$\langle S_z \rangle = \chi_{1/2}^\dagger \hat{S}_z\, \chi_{1/2} = \frac{1}{2}\left(|c_1|^2 - |c_2|^2\right) = \frac{1}{2}\,n_z = \frac{1}{2}\cos\theta_n, \qquad (3.12)$$

where θ_n is the polar angle of the unit vector $\vec{n}$. Thus, the average value $\langle S_z \rangle$ equals to the quasiclassical value of the spin z-component if the spin points in the direction of the unit vector $\vec{n}$. The same conclusion is valid for the x- and y-components of the spin, pointing in the $\vec{n}$-direction:

$$\langle S_x \rangle \; = \chi_{1/2}^\dagger \hat{S}_x \chi_{1/2} = \frac{1}{2}\left(c_1 c_2^* + c_1^* c_2\right) = \frac{1}{2}n_x \; = \frac{1}{2}\sin\theta_n \cos\phi_n,$$

$$(3.13)$$

$$\langle S_y \rangle \; = \chi_{1/2}^\dagger \hat{S}_y \chi_{1/2} = \frac{i}{2}\left(c_1 c_2^* - c_1^* c_2\right) = \frac{1}{2}n_y \; = \frac{1}{2}\sin\theta_n \sin\phi_n,$$

where ϕ_n is the azimuthal angle of the vector $\vec{n}$.

The rotation of the coordinate system by the angle θ about the axis j ($j = x, y, z$) can be described by the unitary rotational operator $\hat{R}_j$:

$$\hat{R}_j = \exp(-i\theta \hat{S}_j) = \hat{E} + (-i\theta \hat{S}_j) + \frac{1}{2!}(-i\theta \hat{S}_j)^2 + \frac{1}{3!}(-i\theta \hat{S}_j)^3 + \dots, \quad (3.14)$$

where $\hat{E}$ is the unit matrix,

$$\hat{E} = \begin{pmatrix} 1 & 0 \\ 0 & 1 \end{pmatrix}. \qquad (3.15)$$

Using the relations, $\hat{S}_x^2 = \hat{S}_y^2 = \hat{S}_z^2 = \frac{1}{4}\hat{E}$, we can rewrite the unitary operator $\hat{R}_j$ in the finite form:

$$\hat{R}_j = \cos\frac{\theta}{2}\,\hat{E} - 2i\sin\frac{\theta}{2}\,\hat{S}_j. \qquad (3.16)$$

We may use the rotational operators in order to transfer, for example, to the S_x-representation. Taking operator $\hat{R}_y$ (rotation about the y-axis) and putting $\theta = \pi/2$, we obtain:

$$\hat{S}'_x = \hat{R}^\dagger_y \, \hat{S}_x \, \hat{R}_y =$$

$$= \frac{1}{2} \begin{pmatrix} \cos\theta/2 & \sin\theta/2 \\ -\sin\theta/2 & \cos\theta/2 \end{pmatrix} \begin{pmatrix} 0 & 1 \\ 1 & 0 \end{pmatrix} \begin{pmatrix} \cos\theta/2 & -\sin\theta/2 \\ \sin\theta/2 & \cos\theta/2 \end{pmatrix} \qquad (3.17)$$

$$= \frac{1}{2} \begin{pmatrix} \sin\theta & \cos\theta \\ \cos\theta & -\sin\theta \end{pmatrix} = \frac{1}{2} \begin{pmatrix} 0 & 1 \\ 0 & -1 \end{pmatrix},$$

where "prime" refers to the S_x-representation. In the same way we may find the operator $\hat{S}'_z$:

$$\hat{S}'_z = \hat{R}^\dagger_y \, \hat{S}_z \, \hat{R}_y = \frac{1}{2} \begin{pmatrix} \cos\theta & -\sin\theta \\ -\sin\theta & -\cos\theta \end{pmatrix} = -\frac{1}{2} \begin{pmatrix} 0 & 1 \\ 1 & 0 \end{pmatrix}. \qquad (3.18)$$

Naturally, the operator $\hat{S}'_x$ in the S_x-representation has the same form as the operator $\hat{S}_z$ in the S_z-representation. The operator $\hat{S}'_z$, within the sign, has the same form as $\hat{S}_x$. The operator $\hat{S}_y$ does not change its form as $\hat{S}_y$ commutes with $\hat{R}_y$.

Next, we consider the spin dynamics for the same cases as in Chap. 2.

1. Larmor precession about the external permanent magnetic field $\vec{B}_{ext} = \{0, 0, B_{ext}\}$. Note that, in fact, we use a "half-quantum" theory as we treat magnetic field classically. The spin Hamiltonian $\mathcal{H}$ has the form:

$$\mathcal{H} = \gamma\hbar\vec{B} \cdot \hat{\vec{S}} = \gamma\hbar B_{ext}\hat{S}_z. \qquad (3.19)$$

The corresponding Schrödinger equation is

$$i\hbar\dot{\chi} = \mathcal{H}\chi. \qquad (3.20)$$

In terms of probability amplitudes c_1 and c_2 we have from (3.4):

$$i\dot{c}_1 = \frac{1}{2}\gamma B_{ext}\, c_1,$$

$$i\dot{c}_2 = -\frac{1}{2}\gamma B_{ext}\, c_2. \tag{3.21}$$

The solution of these equations is obvious:

$$c_1 = c_1(0)\exp(-i\omega_e t/2),$$

$$c_2 = c_2(0)\exp(i\omega_e t/2), \tag{3.22}$$

where $\omega_e = \gamma B_{ext}$ is the ESR frequency.

Note the important quantum effect: the Larmor period $2\pi/\omega_e$ is not the period of the oscillations for the wave function. Indeed, for example, for $t = 0$, the wave function

$$\chi(0) = \begin{pmatrix} c_1(0) \\ c_2(0) \end{pmatrix}, \tag{3.23}$$

and for $t = 2\pi/\omega_e$

$$\chi\left(\frac{2\pi}{\omega_e}\right) = -\begin{pmatrix} c_1(0) \\ c_2(0) \end{pmatrix}. \tag{3.24}$$

The period of motion for the wave function is $4\pi/\omega_e$ instead of $2\pi/\omega_e$. From the other side the average values of the spin components behave like their quasiclassical counterparts. As an example, if $\langle S_y(0)\rangle = 0$, then

$$\langle S_x(t)\rangle = \frac{1}{2}(c_1 c_2^* + c_1^* c_2) = \langle S_x(0)\rangle \cos\omega_e t,$$

$$\langle S_y(t)\rangle = \frac{i}{2}(c_1 c_2^* - c_1^* c_2) = \langle S_x(0)\rangle \sin\omega_e t, \tag{3.25}$$

$$\langle S_z(t)\rangle = \langle S_z(0)\rangle.$$

The z-component of the spin is an integral of motion, as it was in the quasi-classical theory, and the period of precession is $2\pi/\omega_e$.

2. Spin dynamics in the presence of the *rf* field. Taking $\vec{B} = \vec{B}_{ext} + \vec{B}_1$ with $\vec{B}_1 = B_1\{\cos\omega t, \sin\omega t, 0\}$ we start from the following Hamiltonian:

$$\mathcal{H} = \gamma\hbar(B_{ext}\hat{S}_z + B_{1x}\hat{S}_x + B_{1y}\hat{S}_y). \tag{3.26}$$

We will rewrite this Hamiltonian in terms of the operators $\hat{S}_\pm = \hat{S}_x \pm i\hat{S}_y$:

$$\mathcal{H} = \gamma\hbar\left[B_{ext}\hat{S}_z + \frac{1}{2}\left(B_+\hat{S}_- + B_-\hat{S}_+\right)\right], \tag{3.27}$$

where $B_\pm = B_1\exp\pm i\omega t$. It is natural to transfer to the RSC with the x'-axis pointing along the *rf* field $\vec{B}_1$. Using "prime" for the quantities in the RSC we make the following transformation:

$$\mathcal{H}' = \hat{R}_z^\dagger\mathcal{H}\hat{R}_z, \tag{3.28}$$

where

$$\hat{R}_z = \exp(-i\omega t\hat{S}_z) = \cos\frac{\omega t}{2}\hat{E} - 2i\sin\frac{\omega t}{2}\hat{S}_z.$$

Operators $\hat{S}_+$ and $\hat{S}_-$ have a very simple form

$$\hat{S}_+ = \begin{pmatrix} 0 & 1 \\ 0 & 0 \end{pmatrix}, \quad \hat{S}_- = \begin{pmatrix} 0 & 0 \\ 1 & 0 \end{pmatrix}. \tag{3.29}$$

It is easy to check that

$$\hat{S}_z\hat{S}_\pm = \pm\frac{1}{2}\hat{S}_\pm, \quad \hat{S}_\pm\hat{S}_z = \mp\frac{1}{2}\hat{S}_\pm. \tag{3.30}$$

Using these relations we obtain

$$\hat{R}_z^\dagger \hat{S}_\pm \hat{R}_z = \exp(\pm i\omega t)\hat{S}_\pm. \tag{3.31}$$

Thus, the spin Hamiltonian in the RSC takes the form:

$$\mathcal{H}' = \gamma\hbar(B_{ext}\hat{S}_z + B_1\hat{S}_x). \tag{3.32}$$

This Hamiltonian describes the spin, which experiences the constant magnetic field $\vec{B}' = \{B_1, 0, B_{ext}\}$.

In order to write the Schrödinger equation in the RSC we will use the following relations for the wave function χ' in the RSC:

$$\chi' = \hat{R}_z^\dagger \chi,$$

$$\frac{d}{dt}\chi' = \frac{d}{dt}(\hat{R}_z^\dagger)\chi + \hat{R}_z^\dagger\frac{d\chi}{dt}, \tag{3.33}$$

$$\frac{d}{dt}\hat{R}_z^\dagger = (i\omega\hat{S}_z)\hat{R}_z^\dagger.$$

Thus, the Schrödinger equation in the RSC has the form:

$$i\hbar\frac{d}{dt}\chi' + \hbar\omega\hat{S}_z\chi' = \mathcal{H}'\chi', \tag{3.34}$$

or

$$i\hbar\dot{\chi}' = \mathcal{H}'_{eff}\chi' = \left[\gamma\hbar\left(B_{ext} - \frac{\omega}{\gamma}\right)\hat{S}_z + B_1\hat{S}_x\right]\chi'. \tag{3.35}$$

The latter equation describes the spin dynamics in the effective field $\vec{B}_{eff}$, which we have introduced in Chap. 2 (see Eq. (2.19)).

Next, we transfer to another RSC $x''y'z''$ with the z''-axis pointing in the direction of the effective field $\vec{B}_{eff}$. For this transformation we use the unitary operator $\hat{R}_y$, which now describes the rotation about the y'-axis. We have computed already the matrix products $\hat{R}_y^\dagger \hat{S}_x \hat{R}_y$ and $\hat{R}_y^\dagger \hat{S}_z \hat{R}_y$ (see Eqs. (3.17) and (3.18)). Using these expressions and putting

$$\cos\theta = (B_{ext} - \frac{\omega}{\gamma})/B_{eff},$$

$$\sin\theta = B_1/B_{eff}, \tag{3.36}$$

$$B_{eff} = \left[\left(B_{ext} - \frac{\omega}{\gamma} \right)^2 + B_1^2 \right]^{1/2},$$

we obtain the equation for the effective Hamiltonian in our new RSC:

$$\mathcal{H}''_{eff} = \gamma\hbar B_{eff}\hat{S}_z. \tag{3.37}$$

The solution of the Schrödinger equation for this Hamiltonian is given in (3.22), where we have to change ω_e to ω_{eff}. Thus, in the RSC the wave function oscillates with the frequency $\omega_{eff}/2$. In the S''_z-representation connected to the second RSC the operators $\hat{S}_x$, $\hat{S}_y$, and $\hat{S}_z$ describe the spin components along the axes x'', y', z''. Thus, the average spin in the second RSC precesses about the effective field $\vec{B}_{eff}$ like the quasiclassical magnetic moment. The z''-component of the spin, i.e. its components along the effective magnetic field in an integral of motion in our system. Similar analysis can be conducted also for the rotating magnetic field (2.21).

Using the Heisenberg representation generated by the unitary operator $U = \exp(-i\mathcal{H}t/\hbar)$, we can prove that the average spin in any magnetic field evolves like a quasiclassical magnetic moment. In the Heisenberg representation the wave function

$$\chi' = U^\dagger \chi \equiv \chi(0), \tag{3.38}$$

does not change, while the operator, $\hat{S}'_j = U^\dagger \hat{S}'_j U$, evolves according to the Heisenberg equation of motion:

$$\frac{d}{dt}\hat{S}'_j = -\frac{i}{\hbar}[\hat{S}'_j, \mathcal{H}].\tag{3.39}$$

We put $\mathcal{H} = \gamma\hbar\vec{B}\cdot\hat{\vec{S}}'$ and use the commutation relations for the spin operators (3.11). Then, we obtain

$$\frac{d}{dt}\hat{S}'_x = -i\gamma B_y[\hat{S}'_x, \hat{S}'_y] - i\gamma B_z[\hat{S}'_x, \hat{S}'_z] = -\gamma(\hat{S}'_y B_z - \hat{S}'_z B_y).\tag{3.40}$$

In the same way we obtain

$$\frac{d}{dt}\hat{S}'_y = -\gamma(\hat{S}'_z B_x - \hat{S}'_x B_z), \quad \frac{d}{dt}\hat{S}'_z = -\gamma(\hat{S}'_x B_y - \hat{S}'_y B_x).\tag{3.41}$$

These three equations can be written in the vector form:

$$\frac{d}{dt}\hat{\vec{S}}' = -\gamma\hat{\vec{S}}' \times \vec{B}.\tag{3.42}$$

Obviously, the same equation describes the evolution of the average spin $\langle\vec{S}'\rangle$, which does not depend on the representation: $\langle\vec{S}'\rangle = \langle\vec{S}\rangle$. If we multiply the equation for $\langle\vec{S}\rangle$ by $-\gamma$, we will get the quasiclassical equation of motion for the magnetic moment $\vec{\mu}$ (see Eq. (2.3)). Thus, for any magnetic field $\vec{B} = \vec{B}(t)$ the average magnetic moment $-\gamma\langle\vec{S}\rangle$ evolves exactly as the quasiclassical magnetic moment $\vec{\mu}$.

Chapter 4

Mechanical Vibrations of the Cantilever

A cantilever used in MRFM is a tiny beam, which is fixed at one end and free to vibrate at the other end. A small ferromagnetic particle is attached to the cantilever tip (CT). The force produced by a single spin on the ferromagnetic particle affects the parameters of the mechanical vibrations of the CT, which are to be measured in MRFM experiments. The motion of the CT with a ferromagnetic particle with no magnetic force can be described as a simple harmonic motion. The Hamiltonian of the corresponding effective harmonic oscillator can be written in the usual form:

$$\mathcal{H} = -\frac{1}{2m^*}\hat{p}_c^2 + \frac{1}{2}k_c x_c^2, \tag{4.1}$$

where x_c is the coordinate of the CT (i.e. the coordinate of the center of mass of the ferromagnetic particle), m^* is the mass of the effective harmonic oscillator, $\hat{p}_c = -i\hbar\partial/\partial x_c$ is its momentum and k_c is its spring constant. Experimentally k_c can be found using the equipartition theorem

$$\frac{1}{2}k_c\langle x_c^2\rangle = \frac{1}{2}k_B T, \tag{4.2}$$

where $\langle x_c^2\rangle$ is the variance for the thermomechanical random vibrations of the CT and T is the cantilever temperature. The mass m^* can be found from the equation for the experimentally measured oscillator frequency ω_c: $\omega_c^2 = k_c/m^*$. The parameters k_c and m^* can be also computed theoretically using the elasticity theory.

It is convenient to use the operators of annihilation $\hat{a}$ and creation $\hat{a}^\dagger$ in the theory of the harmonic oscillator:

$$\hat{a} = \sqrt{\frac{m^*\omega_c}{2\hbar}} \left(x_c + \frac{i}{m^*\omega_c}\hat{p}_c \right),$$

$$\hat{a}^\dagger = \sqrt{\frac{m^*\omega_c}{2\hbar}} \left(x_c - \frac{i}{m^*\omega_c}\hat{p}_c \right). \tag{4.3}$$

The commutator of these operators is: $[\hat{a}, \hat{a}^\dagger] = 1$. The Hamiltonian of the harmonic oscillator (4.1) can be written in terms of the creation and annihilation operators as:

$$\mathcal{H} = \hbar\omega_c \left(\hat{a}^\dagger\hat{a} + \frac{1}{2} \right). \tag{4.4}$$

The product of the operators $\hat{a}^\dagger\hat{a}$ has the eigenvalues $n = 0, 1, 2, \ldots$. The corresponding eigenfunctions $u_n(x_c)$ can be expressed in terms of the Hermite polynomials H_n:

$$u_n(x_c) = \left(\frac{1}{2^n n!} \right)^{1/2} \left(\frac{m^*\omega_c}{\pi\hbar} \right)^{1/4} \exp\left(-\frac{m^*\omega_c x_c^2}{2\hbar} \right) H_n \left(\sqrt{\frac{m^*\omega_c}{\hbar}} x_c \right),$$

$$H_n(p) = (-1)^n \exp(p^2) \frac{d^n}{dp^n} \left[\exp(-p^2) \right]. \tag{4.5}$$

In Dirac notation the eigenfunctions of the harmonic oscillator Hamiltonian can be denoted as $|n\rangle$. The annihilation operator transforms the eigenstate $|n\rangle$ into the eigenstate $|n-1\rangle$:

$$\hat{a}|n\rangle = \sqrt{n}\,|n-1\rangle,$$

$$\hat{a}|0\rangle = 0. \tag{4.6}$$

The creation operator transforms the state $|n\rangle$ into $|n+1\rangle$:

$$\hat{a}^{\dagger}|n\rangle = \sqrt{n+1}\,|n+1\rangle. \tag{4.7}$$

The eigenfunctions of the operator $\hat{a}$ are called coherent states. An arbitrary complex number α is the eigenvalue of the operator $\hat{a}$. The normalized eigenfunction $u_{\alpha}(x_c)$ corresponding to the eigenvalue α can be written as:

$$u_{\alpha}(x_c) = \left(\frac{m^*\omega_c}{\pi\hbar}\right)^{1/4} \exp\left[-\left(\sqrt{\frac{m^*\omega_c}{2\hbar}}x_c - \alpha\right)^2 + \frac{\alpha^2 - |\alpha|^2}{2}\right]. \tag{4.8}$$

The complex number α is connected to the average values of $\langle x_c\rangle$ and $\langle p_c\rangle$:

$$\alpha = \left(\frac{m^*\omega_c}{2\hbar}\right)^{1/2}\left(\langle x_c\rangle + i\frac{\langle p_c\rangle}{m^*\omega_c}\right). \tag{4.9}$$

The coherent states (4.8) have important properties which allow using them to describe the quasiclassical motion of the CT. First, the variances of the coordinate and momentum of the coherent state are given by:

$$\langle\alpha|(x_c - \langle x_c\rangle)^2|\alpha\rangle = \frac{\hbar}{2m^*\omega_c},$$

$$\langle\alpha|(\hat{p}_c - \langle p_c\rangle)^2|\alpha\rangle = \frac{\hbar m^*\omega_c}{2}, \tag{4.10}$$

where we use the Dirac notation $|\alpha\rangle$ for the state which is described by the eigenfunction $u_{\alpha}(x_c)$, and for any operator $\hat{A}$ we have

$$\langle\alpha|\hat{A}|\alpha\rangle = \int_{-\infty}^{+\infty} u_{\alpha}^*(x_c)\hat{A}\,u_{\alpha}(x_c)\,dx_c.$$

Thus, the product of two variances has the minimum possible value $(\hbar/2)^2$, which does not depend on the value of α. Moreover, in natural quantum units for the length $(\hbar/m^*\omega_c)^{1/2}$ and momentum $(\hbar m^*\omega_c)^{1/2}$ the coordinate and momentum have equal dimensionless uncertainties $1/2$. This is what we may expect for the quasiclassical oscillator.

Second, it follows from the Schrödinger equation, $i\hbar\dot\psi = \mathcal{H}\psi$, that the evolution of the coherent state is given by:

$$\psi(x_c, t) = \left(\frac{m^*\omega_c}{\pi\hbar}\right)^{1/4} \exp\left\{-\frac{i\omega_c t}{2} - \left[\sqrt{\frac{m^*\omega_c}{2\hbar}}x_c - \alpha(t)\right]^2 + \frac{\alpha(t)^2 - |\alpha(t)|^2}{2}\right\},$$

$$\alpha(t) = \alpha(0)\exp(-i\omega_c t), \tag{4.11}$$

where $\psi(x_c, t)$ is the wave function of the harmonic oscillator, $\alpha(0)$ is the value of the parameter α in the initial coherent state which is described by the wave function $\psi(x_c, 0)$. It follows that the harmonic oscillator remains in the coherent state in the process of its motion. The average values of the coordinate and momentum evolve like their classical counterparts:

$$\langle x_c(t)\rangle = \langle x_c(0)\rangle \cos\omega_c t + \frac{\langle p_c(0)\rangle}{m^*\omega_c}\sin\omega_c t,$$

$$\langle p_c(t)\rangle = \langle p_c(0)\rangle \cos\omega_c t - m^*\omega_c\langle x_c(0)\rangle \sin\omega_c t. \tag{4.12}$$

Again, this is what we may expect for the quasiclassical harmonic oscillator. Thus, the coherent states look like a perfect tool for the description of the quasiclassical motion of the CT.

We will also note the formula for the expansion of the coherent state in terms of the eigenstates of the harmonic oscillator Hamiltonian:

$$|\alpha\rangle = \exp(-|\alpha|^2/2)\sum_{n=0}^{\infty}\frac{\alpha^n}{\sqrt{n!}}|n\rangle. \tag{4.13}$$

This formula is convenient for the numerical simulations of the CT motion. The value of $|\alpha|^2$ is equal to the average value of $\hat{n}$: $\langle n\rangle = \langle\alpha|\hat{a}^\dagger\hat{a}|\alpha\rangle = |\alpha|^2$. The quasiclassical motion of the CT corresponds to values $|\alpha| \gg 1$. For $\alpha = 0$ the coherent state coincides with the ground state of the harmonic oscillator.

A spin interacting with the ferromagnetic particle on the CT affects the motion of the CT. In order to write down the operator describing the spin-CT interaction, we will consider the expression for the interaction between the quasiclassical magnetic moment of the spin and the ferromagnetic particle. The magnetic energy U_m of a spin magnetic moment $\vec{\mu}$ in a magnetic field $\vec{B}$ is, as usual, $U_m = -\vec{\mu} \cdot \vec{B}$. The back action of the spin on the CT appears because the magnetic field on the spin depends on the CT coordinate x_c. As a result, the spin exerts a force $\vec{F}$ on the CT, which is equal in magnitude and opposite in direction to the force $\vec{F}'$ produced by the CT on the spin. We assume that the spin is localized, and we ignore the force $\vec{F}'$, while the force $\vec{F}$ is a key quantity in any MRFM technique. To express the magnetic energy U_m in terms of x_c we first approximate the magnetic field near the spin location $\vec{r}_0$ as

$$B_j = B_j(\vec{r}_0) + \frac{\partial B_j}{\partial \vec{r}} \cdot (\vec{r} - \vec{r}_0), \qquad j = x, y, z. \tag{4.14}$$

Thus, the magnetic energy U_m can be written as:

$$U_m = U_0 - \left(\mu_x \frac{\partial B_x}{\partial \vec{r}} + \mu_y \frac{\partial B_y}{\partial \vec{r}} + \mu_z \frac{\partial B_z}{\partial \vec{r}} \right) \cdot (\vec{r} - \vec{r}_0), \tag{4.15}$$

where $U_0 = \vec{\mu} \cdot \vec{B}(\vec{r}_0)$ does not depend on $\vec{r}$. The magnetic force $\vec{F}'$ experienced by the spin is:

$$\vec{F}' = -\frac{\partial U_m}{\partial \vec{r}} = \mu_x \frac{\partial B_x}{\partial \vec{r}} + \mu_y \frac{\partial B_y}{\partial \vec{r}} + \mu_z \frac{\partial B_z}{\partial \vec{r}}. \tag{4.16}$$

The magnetic force experienced by the CT is $\vec{F} = -\vec{F}'$. If, for example, the CT oscillates along the x-axis, the only important component of the force $\vec{F}$ is the x-component:

$$F_x = -\left(\mu_x \frac{\partial B_x}{\partial x} + \mu_y \frac{\partial B_y}{\partial x} + \mu_z \frac{\partial B_z}{\partial x} \right). \tag{4.17}$$

The corresponding energy of the CT is:

$$U_c = \left(\mu_x \frac{\partial B_x}{\partial x} + \mu_y \frac{\partial B_y}{\partial x} + \mu_z \frac{\partial B_z}{\partial x} \right) x_c. \tag{4.18}$$

Note, that using the Maxwell equation $\vec{\nabla} \times \vec{B} = 0$, we can rewrite the expression for the j-th component of the force $\vec{F}'$ as following:

$$F_j' = \mu_x \frac{\partial B_x}{\partial r_j} + \mu_y \frac{\partial B_y}{\partial r_j} + \mu_z \frac{\partial B_z}{\partial r_j} = \mu_x \frac{\partial B_j}{\partial x} + \mu_y \frac{\partial B_j}{\partial y} + \mu_z \frac{\partial B_j}{\partial z}, \qquad (4.19)$$

where $r_x = x$, $r_y = y$, and $r_z = z$. In vector notation the last expression can be written as

$$\vec{F}' = (\vec{\mu} \cdot \vec{\nabla})\vec{B}. \qquad (4.20)$$

Correspondingly, the expression (4.18) for U_c can be rewritten as:

$$U_c = \left(\mu_x \frac{\partial B_x}{\partial x} + \mu_y \frac{\partial B_x}{\partial y} + \mu_z \frac{\partial B_x}{\partial z} \right) x_c = x_c \, (\vec{\mu} \cdot \vec{\nabla}) B_x. \qquad (4.21)$$

Note that the gradient of the magnetic field is taken at the spin location $\vec{r}_0$. Finally, the magnetic energy of the spin-CT system is:

$$U_m = -\vec{\mu} \cdot \vec{B}(\vec{r}_0) + x_c \, (\vec{\mu} \cdot \vec{\nabla}) B_x. \qquad (4.22)$$

The corresponding quantum operator is

$$\hat{U}_m = \gamma \hbar \hat{\vec{S}} \cdot \vec{B}(\vec{r}_0) - \gamma \hbar x_c \, (\hat{\vec{S}} \cdot \vec{\nabla}) B_x. \qquad (4.23)$$

We use this operator when we consider the quantum theory of MRFM.

Next, we will give some basic information about the vibrations of a uniform rectangular cantilever which is fixed at one end. We assume that the length l_c of the cantilever is much greater than its width w_c which, in turn, is much greater than the cantilever thickness $t_c : l_c \gg w_c \gg t_c$. Let the axis of the cantilever be parallel to the x-axis, the cantilever end fixed at $x = 0$, and consider cantilever vibrations in the z-direction (see Fig. 4.1).

The Hamiltonian of the cantilever can be represented in the form

$$\mathcal{H} = \frac{1}{2} \int_0^{l_c} dx \left[\rho S \left(\frac{\partial z_p}{\partial t} \right)^2 + Y I \left(\frac{\partial^2 z_p}{\partial x^2} \right)^2 \right]. \qquad (4.24)$$

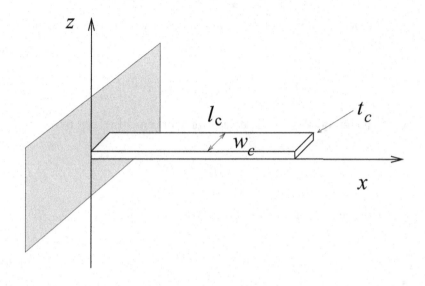

Figure 4.1: Rectangular cantilever with length l_c, width w_c and thickness t_c, fixed at $x = 0$.

Here $z_p = z_p(x,t)$ is the cantilever displacement at a point x, $S = w_c t_c$ is the cross-sectional area of the cantilever, ρ is its density, Y is its Young's modulus, $I = w_c t_c^3/12$. The equation for the cantilever motion with no external force and damping is given by:

$$\rho S \frac{\partial^2 z_p}{\partial t^2} = -Y I \frac{\partial^4 z_p}{\partial x^4}. \tag{4.25}$$

The boundary conditions for the function $z_p(x,t)$ are:

$$z_p(x=0) = \frac{\partial z_p}{\partial x}(x=0) = \frac{\partial^2 z_p}{\partial x^2}(x=l_c) = \frac{\partial^3 z_p}{\partial x^3}(x=l_c) = 0. \tag{4.26}$$

The cantilever eigenfunctions $f_j(x)$ and the eigenfrequencies ω_j, satisfy the equation

$$\rho S \, \omega_j^2 f_j(x) = Y I \frac{\partial^4 f_j(x)}{\partial x^4}. \tag{4.27}$$

The eigenfunctions $f_j(x)$ are orthogonal to each other and can be normalized to the cantilever length l_c:

$$\int_0^{l_c} dx \; f_j(x) f_m(x) = \delta_{jm} \; l_c. \tag{4.28}$$

The lowest eigenfrequency of the cantilever vibrations is given by:

$$\omega_c = \omega_1 \simeq 1.04 \left(\frac{t_c}{l_c^2} \right) \left(\frac{Y}{\rho} \right)^{1/2}. \tag{4.29}$$

The other frequencies can be described by the formula $(j > 1)$:

$$\omega_j \simeq [\pi(j - 0.5)]^2 \left(\frac{t_c}{l_c^2} \right) \left(\frac{Y}{12\rho} \right)^{1/2}. \tag{4.30}$$

The CT amplitude for any mode is twice the amplitude of the mode. As a result, the effective mass $m^* = m_c/4$, where $m_c = \rho l_c \omega_c t_c$ is the cantilever mass.

More detailed informations about the cantilever vibrations can be found in textbooks on the theory of elasticity (e.g. in the book of Landau and Lifshitz [14]).

Chapter 5

Single-Spin Detection in Magnetic Force Microscopy (MFM)

In this and next chapters we consider the theory of a single-spin MFM with no magnetic resonance [15, 16]. While the experimental implementation of a single-spin MFM is unlikely, the theoretical description for MFM is simpler than for MRFM and therefore may help to understand the more complicated MRFM techniques. Below we estimate the static CT displacement caused by a single spin and the decoherence time in MFM.

5.1 Static displacement of the cantilever tip (CT)

We will consider the MFM setup shown in Fig. 5.1. A small ferromagnetic particle having magnetic moment $\vec{m}$ is attached to the CT and interacts with a single spin $\vec{S}$ in a sample. The equilibrium position of the CT must depend on the spin direction. One may expect that the equilibrium position of CT may accept two values corresponding to the two values of the spin z-component $S_z = \pm 1/2$.

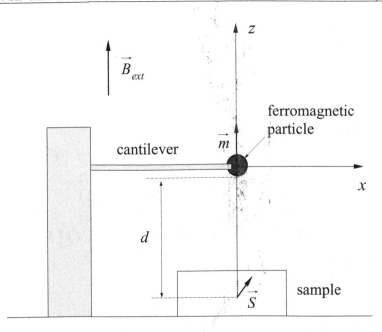

Figure 5.1: MFM setup. $\vec{m}$ is the magnetic moment of the ferromagnetic particle, $\vec{S}$ is a single spin, d is the distance between the bottom of the ferromagnetic particle and the spin.

Let assume that the external permanent magnetic field B_{ext} is large enough: $k_B T \ll 2\mu_B B_{ext}$, where μ_B is the Bohr magneton. Thus, the spin is in its ground state: the spin magnetic moment $\vec{\mu}$ points in the positive z-direction (even without taking into consideration the magnetic field produced by the ferromagnetic particle). We should note that the external magnetic field $\vec{B}_{ext}$ as well as the dipole magnetic field $\vec{B}_d$ produced by the ferromagnetic particle induce a dipole magnetic moment $\vec{m}_d$ in the diamagnetic sample containing a single spin. The polarization of the diamagnetic atoms weakens the attraction between the spin magnetic moment and the CT.

Let the ferromagnetic particle have a spherical shape with a radius $R_0 = 15$ nm. Suppose that the distance d between the particle and the spin, unlike the case shown in Fig. 5.1, is three times less than the radius of the particle: $d = 5$ nm. We take the ratio $R_0/d = 3$ because for the fixed value of d this ratio provides the maximum attraction between the ferromagnetic particle and the magnetic moment of the spin.

The dipole magnetic field produced by the ferromagnetic particle on the spin is

$$B_d = \frac{2}{3}\mu_0 M_0 \left(\frac{R_0}{R_0 + d}\right)^3, \tag{5.1}$$

where $\mu_0 = 4\pi \times 10^{-7} H/m$ is the permeability of the free space, and M_0 is the magnetization (magnetic moment per unit volume) of the ferromagnetic particle. Taking, for example, $\mu_0 M_0 = 1\ T$, we obtain $B_d \approx 0.28\ T$. To

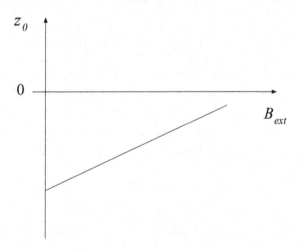

Figure 5.2: An expected dependence of the CT equilibrium position z_0 on the external magnetic field, B_{ext}. The origin is placed at the CT equilibrium position with no spin.

estimate the magnetic moment of the diamagnetic sample effectively interacting with the ferromagnetic particle, we consider a rectangular solid with an area $(2R_0)^2$ and a depth 5.2 nm (the dipole magnetic field, B_d, halves at this depth). Assuming that the magnetic susceptibility of the sample is

3×10^{-7} (as in silicon), we obtain for the magnetic moment of the sample that effectively interacts with the ferromagnetic particle:

$$m_d \approx 3.6 \times 10^{-25} \text{ J/T} \quad \text{for} \quad B_{ext} + B_d = 0.28 \text{ T},$$
$$m_d \approx 6 \times 10^{-24} \text{ J/T} \quad \text{for} \quad B_{ext} + B_d = 5 \text{ T}. \tag{5.2}$$

Changing the external magnetic field from zero to approximately 5 T one can vary the magnetic moment, m_d, from $m_d \ll \mu_B$ up to a value comparable with $\mu_B \approx 9.4 \times 10^{-24}$ J/T. Fig. 5.2 demonstrates the expected dependence of the CT equilibrium position, z_0, on the external magnetic field B_{ext}.

Now we can estimate the value of the static CT displacement, z_0. The magnetic force $\vec{F}$, produced by the spin magnetic moment on CT, points in the negative z-direction and has a magnitude

$$F = \mu_B \frac{\partial B_d}{\partial z} = 2\mu_B \frac{\mu_0 M_0}{R_0 + d} \left(\frac{R_0}{R_0 + d} \right)^2 \approx 390 \text{ aN}. \tag{5.3}$$

The corresponding static displacement of the CT is $z_0 = -F/k_c$, where k_c is the CT spring constant. Taking the value $k_c = 6.5$ μN/m reported by Stowe et al. in Ref. [17], we obtain $z_0 \approx -60$ pm. This value should be compared with the thermal vibrations of the CT. The root-mean-square vibration amplitude at temperature T, can be estimated as

$$z_{rms} \approx \left(\frac{k_B T}{k_c} \right)^{1/2}. \tag{5.4}$$

This is smaller than the displacement of the cantilever, z_0, at temperatures

$$T < F^2/k_B k_c \approx 1.7 \text{ mK}. \tag{5.5}$$

When estimating z_{rms} we assume that the bandwidth of the measuring device, ω_b, is larger than the cantilever frequency, ω_c, as the noise spectral density has a maximum at $\omega = \omega_c$.

A more serious assumption is that the system is in thermodynamic equilibrium. This assumes that we neglect slow relaxation processes that cause the so called "$1/f$ noise". Generally, $1/f$ noise, originated, for example, from the tip-sample electrostatic interaction, can be more important than the thermodynamic noise considered above. More detailed information about the thermal noise one can find, for example, in the book of Kogan[18].

To reduce noise, we can consider the opportunity of decreasing the bandwidth, ω_b, of the measuring device. The price for reducing ω_b is the increase in measurement time. If $\omega_c/Q \ll \omega_b \ll \omega_c$ (where Q is the quality factor of the CT vibrations) one cannot observe oscillations of the CT near its equilibrium position. In this case, one can observe only the relaxation of the cantilever to its equilibrium position.

We would like to note that the MFM setup considered here allows detection of single-spin flips caused by the relaxation process. Let assume that the coercivity of the ferromagnetic particle is larger than the value of the dipole field B_d at the spin. We also assume that a single-spin relaxation time is much larger than the relaxation time (time constant) T_r of the CT vibrations. The CT relaxation time can be found from the formula $T_r = Q/\omega_c$. For the cantilever reported in Ref. [17], the frequency $\omega_c/2\pi = 1.7$ kHz, the quality factor $Q = 6700$, so the relaxation time $T_r = 630$ ms.

Suppose that an experimentalist reverses the direction of the external magnetic field $\vec{B}_{ext}$ and reduces its magnitude to a value less than B_d. In this case the direction of the magnetic moment $\vec{m}$ does not change due to the coercivity of the ferromagnetic particle. Thus, the direction of the dipole field, $\vec{B}_d$, on the spin does not change, either. Next, suppose that the total magnetic field on the spin, $(B_d - B_{ext})$, is reduced to the value which satisfies the inequality:

$$2\mu_B(B_d - B_{ext}) < k_B T. \qquad (5.6)$$

As an example, for $T = 1$ mK the difference $(B_d - B_{ext})$ must be less than or of the order of 200 μT. In this case a single-spin will randomly change its direction. The average time between jumps will determine the spin relaxation time. After each jump, the equilibrium position of the CT changes. Thus, each spin flip generates damped oscillations of the cantilever near the new equilibrium position. In this case, an experimentalist might observe a sequence of short-time CT oscillations such as that shown in Fig. 5.3. If the

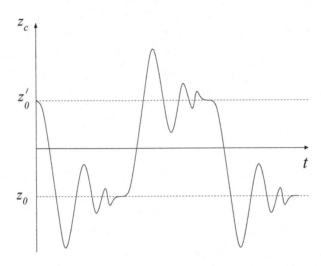

Figure 5.3: Damped oscillations of the CT caused by random jumps of the spin. z_0 and z_0' are the equilibrium positions of the CT for two directions of the spin.

bandwidth, ω_b, of the measuring device is less than ω_c $(\omega_c/Q \ll \omega_b \ll \omega_c)$, one can observe a smooth change of the equilibrium position of the CT with characteristic time $T_r = Q/\omega_c$ rather than the damped oscillations shown in Fig. 5.3. We should note that the experimental observation of the single-spin flips is possible only if the spectral density of this "spin noise" is greater than the spectral density of the $1/f$ noise.

5.2   Decoherence time

Let assume that our spin is placed initially into a superposition of two stationary states: one state corresponds to the positive z-direction, the other corresponds to the negative z-direction (Fig. 5.4). Due to the interaction between the spin and the ferromagnetic particle the cantilever will transform into a superpositional state, which describes two positions (trajectories) of

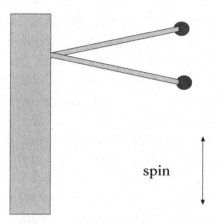

Figure 5.4: The Schrödinger cat state of the spin-cantilever system.

the cantilever at the same time (Fig. 5.4). Since the cantilever is a quasiclassical object, such a state is commonly called a Schrödinger cat state. The considered case is typical in quantum mechanics: a quantum system (spin) interacting with a quasiclassical system (cantilever) generates a Schrödinger cat state of the quasiclassical system.

Typically, the Schrödinger cat state cannot be observed experimentally due to the strong interaction between the quasiclassical system and its environment. This interaction causes the process which is called "decoherence". It means that the coherence (quantum connection) between the two distinguished macroscopic states (two positions of a cantilever) disappears. In other words the Schrödinger cat wave function of the spin-CT system collapses: the cantilever acquires one of the two possible trajectories, and the spin acquires one of the two possible stationary states.

Normally, the decoherence time is very short. That is why the experimental detection of the Schrödinger cat state is so difficult. Below we estimate the decoherence time T_d for the parameters used in Sec. 5.1.

Our rough estimate of the decoherence time is based on the uncertainty relation. We will consider a particle of mass m in a thermal environment at temperature T. Let initially the particle simultaneously occupies two positions separated by the distance Δz (the Schrödinger cat state). We will consider the random motion of the particle along the z-axis. We assume that a Schrödinger cat wave function collapses when the diffusion in the momentum space $\langle \delta p^2(t) \rangle$ becomes close to the "Schrödinger cat momentum uncertainty": $\Delta p^2 \sim (\hbar/\Delta z)^2$.

For a particle interacting with a thermostat, the characteristic fluctuation of energy, δE, during the characteristic time of the fluctuations can be estimated as $k_B T$. If the particle is in equilibrium with the thermostat, the average value of momentum is zero, so $\delta E = \delta p^2/2m$. Thus, $\delta p^2 \sim m k_B T$. The characteristic duration of the particle's fluctuations in the equilibrium position can be estimated as the relaxation time T_r. The diffusion coefficient in the momentum space is $D = \delta p^2/T_r \sim m k_B T/T_r$.

After the creation of a Schrödinger cat state (at $t = 0$), the diffusion, $\langle \delta p^2(t) \rangle$ can be estimated as

$$\langle \delta p^2(t) \rangle \sim Dt \sim \frac{m k_B T}{T_r} t. \tag{5.7}$$

Decoherence occurs when

$$\langle \delta p^2(t) \rangle \approx \Delta p^2 \sim \left(\frac{\hbar}{\Delta z} \right)^2. \tag{5.8}$$

It happens at time

$$t = T_d \sim T_r \frac{\hbar^2}{m k_B T \Delta z^2}. \tag{5.9}$$

Setting the value of separation $\Delta z = 2|z_0| = 1.2 \times 10^{-10}$ m, the mass of the particle $m = m^* = k_c/\omega_c = 6 \times 10^{-14}$ kg, the temperature $T = 1$ mK we obtain from (5.9) $T_d/T_r \sim 9 \times 10^{-10}$. Thus, the decoherence time is negligible compared to the relaxation time. This is a typical situation for quasiclassical systems.

Chapter 6

Transient Process in MFM — The Exact Solution of the Master Equation

Transient process in MFM is one of a few non-trivial problems in the theory of quantum measurement, which allows exact analytical solution. In this chapter we will derive and discuss the analytical solution for the MFM transient dynamics.

6.1 Hamiltonian and master equation for the spin-CT system

In mechanics the transient process for a single harmonic oscillator under the action of a constant external force is described as damped oscillations with the frequency $\omega_c(1 - 1/4Q^2)^{1/2}$ and the relaxation time (time constant) $T_r = Q/\omega_c$. In order to describe the thermal fluctuations of the oscillator one has to consider the classical distribution function, in our case $f(z_c, v_c, t)$. The distribution function describes the probability for the oscillator to have the coordinate z_c and the velocity v_c at the time t. If the oscillator starts from a fixed initial position and velocity then the probability function eventu-

41

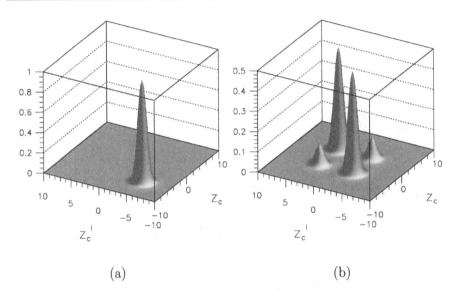

(a)   (b)

Figure 6.1: Density matrix modulus : (a) for the coherent quasiclassical state, (b) for the Schrödinger cat state.

ally spreads due to the thermal fluctuations. This process is called thermal diffusion. For the ensemble of identical oscillators starting from the same initial conditions the distribution function describes the real distribution of oscillators in the plane $z_c - v_c$.

In quantum mechanics, instead of the distribution function, one can use the density matrix $\rho(z_c, z'_c, t)$. If an oscillator is in the coherent quasiclassical state its density matrix is given by

$$\rho(z_c, z'_c, t) = \psi(z_c, t)\, \psi^*(z'_c, t), \tag{6.1}$$

where the wave function ψ of the coherent state is defined in (4.11). The diagonal elements of the density matrix $\rho(z_c, z_c, t)$ describe the probability density to find the oscillator at the point z_c. The density matrix modulus $|\rho|$ in (6.1) describes roughly the narrow peak which oscillates along the diagonal $z_c = z'_c$ in the $z_c - z'_c$ plane (see Fig. 6.1(a)). If the oscillator is in the superposition of two coherent quasiclassical states (the Schrödinger cat state) then the density matrix modulus describes roughly two diagonal peaks, which correspond to two positions of the cantilever (see Fig. 6.1(b)). Besides, the density matrix describes the two off-diagonal peaks. These peaks reflect the coherence between the two oscillator positions. They signal that the oscillator occupies two distinct positions at the same time.

The interaction with the thermal environment causes the decoherence of the Schrödinger cat state: the off-diagonal peaks in Fig. 6.1(b) disappear. It means that the Schrödinger cat wave function collapses, and the oscillator chooses one of the two possible trajectories. (Fig. 6.1(b)) corresponds to the equal probabilities for two trajectories.) In the ensemble of identical oscillators half of them "chooses" the first trajectory and the other half chooses the second trajectory. The classical effect of thermal diffusion will cause the stretching of the diagonal peaks along the $z = z'$ diagonal. Finally, at times $t \gg T_r$ the density matrix will describe a single peak centered at the oscillator equilibrium position (the origin) and stretched along the diagonal by the value $(k_B T / k_c)^{1/2}$ which corresponds to the thermal vibrations of the oscillator.

The equation which describes the evolution of the density matrix in the thermal environment is called the master equation. The effect of the environment depends on its "spectral density", i.e. the density of environmental oscillators at a given frequency ω. Probably the simplest model of the environment is the "ohmic" model, where the spectral density is proportional to the frequency ω for $\omega < \Omega$, where Ω is the cutoff frequency for the environment.

For the ohmic model, the simplest master equation has been obtained by Caldeira and Leggett [19]. This equation is valid in the high temperature limit $k_B T \gg \hbar \Omega$. The master equation derived by Unruh and Zurek [20] is valid for an arbitrary temperature. Hu et al. [21] showed that both equations ([19] and [20]) fail at times shorter than or close to $\hbar / k_B T$, and derived the master equation for the non-ohmic environment. In our book we will use the simplest master equation by Caldeira and Leggett [19].

We will consider the same setup as in the previous chapter (see Fig. 5.1).

For numerical estimations we will use the same values of parameters as in the previous chapter. The density matrix of the spin-cantilever (spin-CT) system will be function of z_c, z'_c, s, s' and t, where the variables $s = S_z$ ans $s' = S'_z$ take the values $\pm 1/2$. We will take into account the interaction between the cantilever and its environment, and ignore the direct interaction between the spin and its environment.

The Hamiltonian of the spin-CT system is the sum of the oscillator Hamiltonian (4.1), of the spin Hamiltonian in the external magnetic field (3.19), and the spin-CT interaction (4.23). For the setup shown in Fig. 5.1, the Hamiltonian takes the form:

$$\mathcal{H} = \frac{\hat{p_c}^2}{2m^*} + \frac{1}{2}k_c z_c^2 + \gamma\hbar B_0 \hat{S}_z - \gamma\hbar\frac{\partial B_d}{\partial z}\hat{S}_z z_c. \tag{6.2}$$

Here $B_0 = B_{ext} + B_d^{(0)}$ is the total magnetic field on the spin when $z_c = 0$, the origin corresponds to the equilibrium position of the CT with no spin, the gradient of the dipole field $\partial B_d/\partial z$ is taken at the spin location when $z_c = 0$. We may eliminate the third term in the Hamiltonian (6.2) transferring to the RSC rotating with the frequency $\gamma\hbar B_0$ (see Chap. 2). Next, we will use the dimensionless Hamiltonian choosing the natural quantum units of length $(\hbar\omega_c/k_c)^{1/2}$, energy $\hbar\omega_c$, momentum $(\hbar\omega_c m^*)^{1/2}$, and force $(\hbar\omega_c k_c)^{1/2}$. Using the same notation for the coordinate and momentum as before we obtain:

$$\mathcal{H} = \frac{1}{2}(\hat{p_c}^2 + z_c^2) - 2\eta z_c \hat{S}_z, \tag{6.3}$$

where the parameter of the spin-CT interaction η is equal to the magnetic force F (5.3) produced by the spin on the CT in the units of $(\hbar\omega_c k_c)^{1/2}$:

$$\eta = \frac{\mu_B}{\sqrt{\hbar\omega_c k_c}}\frac{\partial B_d}{\partial z}. \tag{6.4}$$

Using the dimensionless time $\tau = \omega_c t$ we will write the master equation for the spin-CT system

$$\frac{\partial \rho_{s,s'}}{\partial \tau} = \left[\frac{i}{2} (\partial_{zz} - \partial_{z'z'}) - \frac{i}{2}(z_c^2 - z_c'^2) - \frac{1}{2}\beta(z_c - z_c')(\partial_z - \partial_{z'}) \right.$$

$$\left. - D\beta(z_c - z_c')^2 - 2i\eta(z_c's' - z_c s) \right] \rho_{s,s'}.$$

(6.5)

Here $\beta = 1/Q$, $D = k_B T/\hbar\omega_c$, $\partial_{zz} = \partial^2/\partial z_c^2$, $\partial_z = \partial/\partial z_c$. Using new coordinates

$$r = z_c - z_c', \qquad R = \frac{1}{2}(z_c + z_c'), \qquad (6.6)$$

Eq. (6.5) can be written as:

$$\frac{\partial \rho_{s,s'}(R, r, \tau)}{\partial \tau} =$$

$$\left\{ i\partial_{Rr} - iRr - \beta r\partial_r - D\beta r^2 - i\eta \left[(2R - r)s' - (2R + r)s \right] \right\} \rho_{s,s'}(R, r, \tau).$$

(6.7)

Performing a Fourier transformation of this equation with respect to the variable "R", one obtains, after re-arrangements,

$$\frac{\partial \hat{\rho}_{s,s'}}{\partial \tau} + (\beta r - k)\frac{\partial \hat{\rho}_{s,s'}}{\partial r} + \left[r + 2\eta(s' - s) \right]\frac{\partial \hat{\rho}_{s,s'}}{\partial k} = \left[-D\beta r^2 + i\eta r(s' + s) \right] \hat{\rho}_{s,s'},$$

(6.8)

where

$$\hat{\rho}_{s,s'}(k, r, \tau) = \int_{-\infty}^{+\infty} e^{ikR} \rho_{s,s'}(R, r, \tau) \, dR \qquad (6.9)$$

We can study separately the spin diagonal case ($s = s'$) and the off-diagonal case ($s \neq s'$). For $s' = s$ (up-up or down-down spins), we have the following equation:

$$\frac{\partial \hat{\rho}_{s,s}}{\partial \tau} + (\beta r - k)\frac{\partial \hat{\rho}_{s,s}}{\partial r} + r\frac{\partial \hat{\rho}_{s,s}}{\partial k} = \left(-D\beta r^2 + 2i\eta rs \right) \hat{\rho}_{s,s}, \qquad (6.10)$$

and for $s' \neq s$ (up-down or down-up spins):

$$\frac{\partial \hat{\rho}_{s,-s}}{\partial \tau} + (\beta r - k)\frac{\partial \hat{\rho}_{s,-s}}{\partial r} + (r + 4\eta s)\frac{\partial \hat{\rho}_{s,-s}}{\partial k} = -D\beta r^2 \, \hat{\rho}_{s,-s}. \qquad (6.11)$$

We will derive the exact solution of the master equation (6.5) for the case when the spin is "prepared" initially in the superposition of two states with $s = 1/2$ and $s = -1/2$, while the CT is in the quasiclassical coherent state

$$\psi(z_c, s, 0) = \frac{1}{(\pi)^{1/4}} \exp\left[i\langle p_c(0)\rangle z_c - \frac{1}{2}(z_c - \langle z_c(0)\rangle)^2\right] \cdot \begin{pmatrix} a \\ b \end{pmatrix}, \qquad (6.12)$$

where the amplitudes a and b correspond to the values of $s = 1/2$ and $s = -1/2$, respectively. The corresponding density matrix can be written as

$$\rho_{s,s'}(z_c, z_c', 0) = \psi(z_c, s, 0) \otimes \psi^\dagger(z_c', s', 0), \qquad (6.13)$$

where we use notation $\otimes$ for the tensor product.

Note, that we consider here an ensemble of spin-CT systems with the same initial state. This implies that the experimenter can detect the position and momentum of a point on the CT with quantum limit accuracy $\langle (\delta p_c)^2 \, (\delta z_c)^2 \rangle = 1/4$. (In our gedanken experiment, this corresponds to an uncertainty of 300 fm for position and 300 nm/s for velocity.) Based on the master equation, we can predict the average position of the CT for its given initial state, depending on the spin state. If the double uncertainty of the position is smaller than the separation between two possible average positions, the cantilever will measure the state of the spin.

After Fourier transformation, the "cantilever part" of the initial density matrix is represented by

$$\hat{\rho}_{s,s'}(k, r, 0) \propto \exp\left[i\langle p_c(0)\rangle r + ik\langle z_c(0)\rangle - r^2/4 - k^2/4\right]. \qquad (6.14)$$

6.2 Solution for spin diagonal matrix elements

The equations for the characteristics of Eq. (6.10) are

$$d\tau = \frac{dr}{\beta r - k} = \frac{dk}{r} = \frac{d\hat{\rho}_{s,s}}{\left(-D\beta r^2 + 2i\eta sr\right)\hat{\rho}_{s,s}}, \tag{6.15}$$

or, explicitly

$$\frac{dr}{d\tau} = \beta r - k,$$

$$\frac{dk}{d\tau} = r, \tag{6.16}$$

$$\frac{d\hat{\rho}_{s,s}}{d\tau} = \left(-D\beta r^2 + 2i\eta sr\right)\hat{\rho}_{s,s}.$$

From the first two equations in (6.16), one obtains

$$\frac{d^2k}{d\tau^2} - \beta\frac{dk}{d\tau} + k = 0,$$

which has the following general solution

$$k = e^{\beta\tau/2}\left(c_1\cos\theta\tau + c_2\sin\theta\tau\right), \tag{6.17}$$

where $\theta = \sqrt{1 - \frac{\beta^2}{4}}$. Here we are considering the case $\beta < 2$, so θ is a real number. Using the second equation in (6.16) one obtains:

$$r = e^{\beta\tau/2}\left[\left(\frac{\beta}{2}\cos\theta\tau - \theta\sin\theta\tau\right)c_1 + \left(\frac{\beta}{2}\sin\theta\tau + \theta\cos\theta\tau\right)c_2\right]. \tag{6.18}$$

Inverting Eqs. (6.17) and (6.18) as functions of c_1 and c_2 one obtains the characteristic curves:

$$c_1 = e^{-\beta\tau/2}(q_1k + q_2r), \tag{6.19}$$

and

$$c_2 = e^{-\beta\tau/2}(p_1k + p_2r), \tag{6.20}$$

where the time dependent constants q_1, q_2, p_1 and p_2 have been defined as

$$q_1 = \frac{1}{\theta}\left(\frac{\beta}{2}\sin\theta\tau + \theta\cos\theta\tau\right),$$

$$q_2 = -\frac{1}{\theta}\sin\theta\tau,$$

$$p_1 = \frac{1}{\theta}\left(-\frac{\beta}{2}\cos\theta\tau + \theta\sin\theta\tau\right),$$

$$p_2 = \frac{1}{\theta}\cos\theta\tau.$$

(6.21)

Substituting (6.18) into the third equation of (6.16) and integrating in time, one obtains:

$$\hat{\rho}_{s,s}(k,r,\tau) \propto A(c_1,c_2)\exp\left[i2\eta s(c_1 g_1 + c_2 g_2) - D\beta(c_1^2 f_1 + 2c_1 c_2 f_3 + c_2^2 f_2)\right],$$

(6.22)

where the functions $f_i's$ and $g_i's$ are defined as

$$f_1(\tau) = \frac{e^{\beta\tau}}{8}\left[\left(\beta + \frac{4\theta^2}{\beta}\right) + \beta\cos 2\theta\tau - 2\theta\sin 2\theta\tau\right],$$

$$f_2(\tau) = \frac{e^{\beta\tau}}{8}\left[\left(\beta + \frac{4\theta^2}{\beta}\right) - \beta\cos 2\theta\tau + 2\theta\sin 2\theta\tau\right],$$

$$f_3(\tau) = \frac{e^{\beta\tau}}{8}\left[2\theta\cos 2\theta\tau + \beta\sin 2\theta\tau\right],$$

$$g_1(\tau) = e^{\beta\tau/2}\cos\theta\tau,$$

$$g_2(\tau) = e^{\beta\tau/2}\sin\theta\tau.$$

(6.23)

The arbitrary function $A(c_1,c_2)$, which depends on the characteristics, is determined by the initial density matrix $\hat{\rho}_{s,s}(k(0), r(0), 0)$,

$$A(c_1, c_2) = \hat{\rho}_{s,s}\left(c_1, \frac{1}{2}\beta c_1 + \theta c_2, 0\right) \exp\left[-2i\eta s(c_1 g_{10} + c_2 g_{20})\right]$$

$$\times \exp\left[D\beta(c_1^2 f_{10} + 2c_1 c_2 f_{30} + c_2^2 f_{20})\right],$$

(6.24)

where $f_{i0} = f_i(0)$ and $g_{i0} = g_i(0)$. From the initial density matrix (Eq. (6.14)), we obtain

$$\hat{\rho}_{s,s}(k, r, 0) \propto \exp\left\{i\left[\left(\frac{1}{2}\langle p_c(0)\rangle\beta + \langle z_c(0)\rangle\right)c_1 + \langle p_c(0)\rangle\theta c_2\right]\right\}$$

$$\times \exp\left\{-\left[\left(\frac{\beta^2}{16} + \frac{1}{4}\right)c_1^2 + \frac{\beta\theta}{4}c_1 c_2 + \frac{\theta^2}{4}c_2^2\right]\right\}.$$

(6.25)

Substituting (6.24) and (6.25) into (6.22), one obtains:

$$\hat{\rho}_{s,s}(k, r, \tau) \propto$$

$$\exp\left\{i\left[\left(\frac{1}{2}\langle p_c(0)\rangle\beta + \langle z_c(0)\rangle + 2\eta s G_1\right)c_1 + \left(\langle p_c(0)\rangle\theta + 2\eta s G_2\right)c_2\right]\right\}$$

$$\times \exp\left\{-\left[\left(\frac{\beta^2}{16} + \frac{1}{4}\right)c_1^2 + \frac{\beta\theta}{4}c_1 c_2 + \frac{\theta^2}{4}c_2^2\right]\right\}$$

$$\times \exp\left\{-D\beta(F_1 c_1^2 + 2c_1 c_2 F_3 + F_2 c_2^2)\right\},$$

(6.26)

where F_i and G_i are defined as

$$F_i(\tau) = f_i(\tau) - f_{i0}, \qquad G_i(\tau) = g_i(\tau) - g_{i0}.$$

Substituting in (6.26), the values of characteristics as functions of k and r (Eqs. (6.19) and (6.20)), one obtains:

$$\hat{\rho}_{s,s}(k, r, \tau) \propto \exp\left[-r^2 C_1 + irC_2 + (iB_2 - rB_1)k - \sigma_*^2 k^2\right], \qquad (6.27)$$

where

$$\sigma_*^2 = e^{-\beta t}\left[\left(\frac{\beta^2}{16}+\frac{1}{4}\right)q_1^2 + \frac{\beta\theta}{4}q_1p_1 + \frac{\theta^2}{4}p_1^2 \quad + D\beta(F_1q_1^2 + 2q_1p_1F_3 + F_2p_1^2)\right],$$

(6.28)

$$B_1 = e^{-\beta t}\left\{\left(\frac{\beta^2}{16}+\frac{1}{4}\right)2q_1q_2 + \frac{\beta\theta}{4}(q_1p_2 + q_2p_1) + \frac{\theta^2}{4}\,2p_1p_2\right.$$

(6.29)

$$\left. +2D\beta[F_1q_1q_2 + (q_1p_2 + q_2p_1)F_3 + F_2p_1p_2]\right\},$$

$$B_2 = e^{-\beta t/2}\left[\left(\frac{1}{2}\langle p_c(0)\rangle\beta + \langle z_c(0)\rangle + 2\eta sG_1\right)q_1 + (\langle p_c(0)\rangle\theta + 2\eta sG_2)\,p_1\right],$$

(6.30)

$$C_1 = e^{-\beta t}\left[\left(\frac{\beta^2}{16}+\frac{1}{4}\right)q_2^2 + \frac{\beta\theta}{4}q_2p_2 + \frac{\theta^2}{4}p_2^2 + D\beta(F_1q_2^2 + 2q_2p_2F_3 + F_2p_2^2)\right], \quad (6.31)$$

$$C_2 = e^{-\beta t/2}\left[\left(\frac{1}{2}\langle p_c(0)\rangle\beta + \langle z_c(0)\rangle + 2\eta sG_1\right)q_2 + (\langle p_c(0)\rangle\theta + 2\eta sG_2)\,p_2\right].$$

(6.32)

Note that in Eqs. (6.30) and (6.32), the coefficients B_2 and C_2 depend on s. Performing the inverse Fourier transform one obtains

$$\rho_{1/2,1/2}(R,r,\tau) = \frac{|a|^2}{2\sqrt{\pi}\,\sigma_*}\exp\left[-r^2C_1 + irC_2(1/2)\right]$$
$$\times \exp\left[(-rB_1 + iB_2(1/2) - iR)^2/4\sigma_*^2\right],$$

$$\rho_{-1/2,-1/2}(R,r,\tau) = \frac{|b|^2}{2\sqrt{\pi}\,\sigma_*}\exp\left[-r^2C_1 + irC_2(-1/2)\right]$$
$$\times \exp\left[(-rB_1 + iB_2(-1/2) - iR)^2/4\sigma_*^2\right].$$

(6.33)

Equations (6.33) represent two squeezed Gaussians with modulus

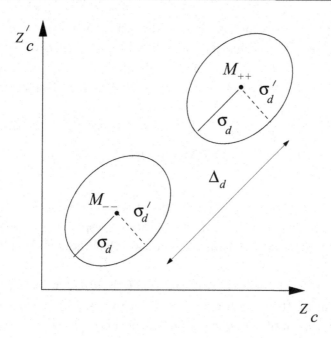

Figure 6.2: Schematic view of the Gaussians representing the diagonal elements $|\rho_{-1/2,-1/2}|$ and $|\rho_{1/2,1/2}|$ (seen from the top) in the (z_c, z_c') plane. We show the centers M_{--} and M_{++}, the variances σ_d' (transverse) and σ_d (parallel), and the distance between the centers Δ_d.

$$|\rho_{1/2,1/2}(R, r, \tau)| = \frac{|a|^2}{2\sqrt{\pi}\,\sigma_*} \exp\left[-r^2(C_1 - B_1^2/4\sigma_*^2)\right]$$
$$\times \exp\left[-(B_2(1/2) - R)^2/4\sigma_*^2\right],$$

$$(6.34)$$

$$|\rho_{-1/2,-1/2}(R, r, \tau)| = \frac{|b|^2}{2\sqrt{\pi}\,\sigma_*} \exp\left[-r^2(C_1 - B_1^2/4\sigma_*^2)\right]$$
$$\times \exp\left[-(B_2(-1/2) - R)^2/4\sigma_*^2\right].$$

Figure 6.2 shows schematically two peaks (seen from the top as ellipses) corresponding to the two matrix elements $|\rho_{1/2,1/2}|$ and $|\rho_{-1/2,-1/2}|$. We denote the centers of the ellipses, which lie on the diagonal $z = z'$, by M_{++}

and M_{--}, the semi-major axis by σ_d, the semi-minor axis by σ'_d and the distance between the centers by Δ_d. The $|\rho^{max}_{1/2,1/2}|$ is located at $M_{++} = (r = 0, R = B_2(1/2))$ or $z_c = z'_c = B_2(1/2)$, while the $|\rho^{max}_{-1/2,-1/2}|$ is at $M_{--} = (r = 0, R = B_2(-1/2))$ or $(z_c = z'_c = B_2(-1/2))$. The distance Δ_d is given by

$$\Delta_d = B_2(1/2) - B_2(-1/2). \tag{6.35}$$

From Eqs. (6.34), we obtain $\sigma_d = \sqrt{2}\,\sigma_*$, and

$$2\sigma'_d{}^2 = \frac{4\sigma_*^2}{4\sigma_*^2 C_1 - B_1^2}. \tag{6.36}$$

For a single spin measurement, the two peaks corresponding to $|\rho^{max}_{1/2,1/2}|$ and $|\rho^{max}_{-1/2,-1/2}|$ must be well separated. It follows that the condition $\Delta_d > 2\sigma_d$ must be satisfied.

First, we consider the case $\beta\tau \gg 1$ or $t \gg Q/\omega_c$, where Q/ω_c is the time constant for the CT vibrations. In this case, we obtain two equilibrium positions for the cantilever, when the transient process is over. We have $\Delta_d = 2\eta$ and $\sigma_d = \sqrt{D}$. The value $\sigma_d = \sqrt{D}$ is the thermodynamical uncertainty in the CT position caused by the thermal noise. The two equilibrium positions can be distinguished if $\eta > \sqrt{D}$ or $T < F^2/k_B k_c$, where $F = \mu_B |\partial B_d/\partial z|$ is the magnetostatic force between the ferromagnetic particle and the spin. The inequality for temperature T exactly coincides with formula (5.5).

Next, we consider the initial transient process after the instant $t = 0$. For $\beta\tau \ll 1$, we have

$$\Delta_d = 4\eta \sin^2 \frac{\tau}{2}, \quad \sigma_d = \left[\frac{1}{2} + D\beta\tau - D\beta \cos\tau \sin^3\tau\right]^{1/2}. \tag{6.37}$$

The expression for Δ_d describes the oscillating distance between the two peaks. It corresponds to initial vibrations of two classical oscillators near their equilibrium positions $z_c = \eta$ and $z_c = -\eta$. The distance between them is given by Δ_d. For our gedanken experiment the maximum value of Δ_d is 0.24 nm. The formula for σ_d contains three terms. The first term, $1/2$, corresponds to the quantum dispersion of the initial wave function. The second term, $D\beta\tau$, describes the initial thermal diffusion of an ensemble of oscillators. Formally, setting $\tau \sim 1/\beta$, we can estimate the final dispersion $\sigma_d = \sqrt{D}$, which corresponds to thermal vibrations of the cantilever. The third term describes insignificant oscillations with small amplitude, $D\beta$.

Note, that the condition for distinguishing two cantilever positions at the beginning of the transient process is much less restrictive than the corresponding condition for the equilibrium positions at $\beta\tau \gg 1$. Indeed, after the first half-period ($\tau = \pi$), we have $\Delta_d = 4\eta$ and $\sigma_d = (1/2 + \pi D\beta)^{1/2}$. Taking into account that $\beta \ll 1$, the condition for distinguishing two positions, $\eta > (1/2 + \pi D\beta)^{1/2}$, is much easier than $\eta > \sqrt{D}$. In our gedanken experiment the condition for distinguishing the two positions for the transient process is

$$T < T_{max} = \frac{4}{\pi} \frac{QF^2}{k_B k_c} = 14 \text{ K}. \tag{6.38}$$

This estimate seems to be too optimistic. It is connected with the very small distance (5 nm) between the ferromagnetic particle and the spin. If we increase this distance to 50 nm, the temperature T_{max} drops from 14 K to 1.1 mK.

The condition $\Delta_d > 2\sigma_d$ is satisfied for the first time at

$$\tau = \tau_0 \approx \frac{2^{1/4}}{\sqrt{\eta}}. \tag{6.39}$$

This expression is valid if $\eta \gg 1$ and $\eta \gg (D\beta)^2/\sqrt{8}$. For our parameters we have $\eta = 144$, $D = 1.25 \times 10^7 T$ (T is the temperature in Kelvin), $\beta = 1.5 \times 10^{-4}$, and $\tau_0 = 0.1$. Thus, the above conditions are both satisfied. The value of $t_0 = \omega_c \tau_0$ is approximately 9.3 μs.

6.3 Solution for spin off-diagonal matrix elements

The equations for the characteristics are now given by

$$d\tau = \frac{dr}{\beta r - k} = \frac{dk}{r - 4\eta s} = \frac{d\hat{\rho}_{s,-s}}{-D\beta r^2 \, \hat{\rho}_{s,-s}}, \tag{6.40}$$

or

$$\frac{dr}{d\tau} = \beta r - k,$$

$$\frac{dk}{d\tau} = r - 4\eta s, \tag{6.41}$$

$$\frac{d\hat{\rho}_{s,-s}}{d\tau} = -D\beta r^2 \, \hat{\rho}_{s,-s}.$$

The solutions of the first two equations of (6.41) are

$$k = e^{\beta\tau/2}\left(c_1 \cos\theta\tau + c_2 \sin\theta\tau\right) + 4\beta\eta s,$$

$$r = e^{\beta\tau/2}\left[\left(\frac{\beta}{2}\cos\theta\tau - \theta\sin\theta\tau\right)c_1 + \left(\frac{\beta}{2}\sin\theta\tau + \theta\cos\theta\tau\right)c_2\right] + 4\eta s.$$
$$\tag{6.42}$$

Following the same steps as above we obtain for the Fourier transform:

$$\hat{\rho}_{1/2,-1/2}(k, r, \tau) \propto A(c_1, c_2)\exp\left\{-D\beta\left[f_1 c_1^2 + 2c_1 c_2 f_3 + f_2 c_2^2\right]\right\} \tag{6.43}$$

$$\times \exp\left\{-D\beta\left[4\eta(g_1 c_1 + g_2 c_2) + 4\eta^2\tau\right]\right\},$$

where we fix $s = 1/2$. (Changing sign of s corresponds to a change of sign of η, see Eqs. (6.41), therefore the case $s = -1/2$ can be easily obtained.)

The functions f_i and g_i have been defined as above, and c_1 and c_2 are new characteristic curves given by

$$c_1 = e^{-\beta\tau/2}(q_1 k + q_2 r + \eta q_3),$$

$$c_2 = e^{-\beta\tau/2}(p_1 k + p_2 r + \eta p_3). \tag{6.44}$$

Here, q_1, q_2, p_1, p_2 are defined by Eqs. (6.21) and q_3, p_3 are given by

$$q_3 = \frac{2}{\theta}\left[-\beta\left(\frac{\beta}{2}\sin\theta\tau + \theta\cos\theta\tau\right) + \sin\theta\tau\right],$$

$$p_3 = \frac{2}{\theta}\left[\beta\left(\frac{\beta}{2}\cos\theta\tau - \theta\sin\theta\tau\right) - \cos\theta\tau\right]. \tag{6.45}$$

With the same initial condition, Eq. (6.14), we can determine the function $A(c_1, c_2)$ and obtain

$$\hat{\rho}_{1/2,-1/2}(k,r,\tau) \propto \hat{\rho}_{1/2,-1/2}\left(c_1 + 2\beta\eta, \frac{1}{2}\beta c_1 + \theta c_2 + 2\eta, 0\right)$$

$$\times \exp\left\{-D\beta\left[F_1 c_1^2 + 2c_1 c_2 F_3 + F_2 c_2^2 + 4\eta(G_1 c_1 + G_2 c_2) + 4\eta^2\tau\right]\right\}, \tag{6.46}$$

where $F_i(\tau)$ and $G_i(\tau)$ are defined as above. By substituting the initial condition (6.14), we have

$$\hat{\rho}_{1/2,-1/2}(k,r,\tau) \propto \exp\left[-r^2 C_{12} - r\eta C_{11} - \eta^2 C_{10} + irC_{21} + i\eta C_{20}\right]$$

$$\times \exp\left[(iB_{20} - rB_{11} - \eta B_{10})k - \sigma_*^2 k^2\right], \tag{6.47}$$

where σ_* is given by Eq. (6.28) and

$$C_{12} = e^{-\beta\tau}\left[\left(\frac{\beta^2}{16} + \frac{1}{4} + D\beta F_1\right)q_2^2 + \left(\frac{\beta\theta}{4} + 2D\beta F_3\right)q_2 p_2 + \left(\frac{\theta^2}{4} + D\beta F_2\right)p_2^2\right],$$

$$C_{11} = e^{-\beta\tau}\left[\left(\frac{\beta^2}{16} + \frac{1}{4} + D\beta F_1\right)2q_2q_3 + \left(\frac{\beta\theta}{4} + 2D\beta F_3\right)(q_2p_3 + p_2q_3)\right]$$

$$+e^{-\beta\tau}\left[\left(\frac{\theta^2}{4} + D\beta F_2\right)2p_2p_3\right] + 4e^{-\beta\tau/2}\left[\left(\frac{3\beta}{8} + D\beta G_1\right)q_2 + \left(\frac{\theta}{4} + D\beta G_2\right)p_2\right],$$

$$C_{10} = e^{-\beta\tau}\left[\left(\frac{\beta^2}{16} + \frac{1}{4} + D\beta F_1\right)q_3^2 + \left(\frac{\beta\theta}{4} + 2D\beta F_3\right)q_3p_3 + \left(\frac{\theta^2}{4} + D\beta F_2\right)p_3^2\right]$$

$$+4e^{-\beta\tau/2}\left[\left(\frac{3\beta}{8} + D\beta G_1\right)q_3 + \left(\frac{\theta}{4} + D\beta G_2\right)p_3\right] + 4\left(\frac{1}{4} + \frac{\beta^2}{4} + D\beta\tau\right),$$

$$C_{21} = e^{-\beta\tau/2}\left[\left(\frac{\langle p_c(0)\rangle\beta}{2} + \langle z_c(0)\rangle\right)q_2 + \langle p_c(0)\rangle\theta p_2\right],$$

$$C_{20} = e^{-\beta\tau/2}\left[\left(\frac{\langle p_c(0)\rangle\beta}{2} + \langle z_c(0)\rangle\right)q_3 + \langle p_c(0)\rangle\theta p_3\right] + 2\left(\langle p_c(0)\rangle + \langle z_c(0)\rangle\beta\right),$$

$$B_{11} = e^{-\beta\tau}\left[\left(\frac{\beta^2}{16} + \frac{1}{4} + D\beta F_1\right)2q_2q_1 + \left(\frac{\beta\theta}{4} + 2D\beta F_3\right)(q_1p_2 + q_2p_1)\right]$$

$$+e^{-\beta\tau}\left[\left(\frac{\theta^2}{4} + D\beta F_2\right)2p_2p_1\right],$$

$$B_{10} = e^{-\beta\tau}\left[\left(\frac{\beta^2}{16} + \frac{1}{4} + D\beta F_1\right)2q_3q_1 + \left(\frac{\beta\theta}{4} + 2D\beta F_3\right)(q_1p_3 + q_3p_1)\right] +$$

$$e^{-\beta\tau}\left[\left(\frac{\theta^2}{4} + D\beta F_2\right)2p_3p_1\right] + 4e^{-\beta\tau/2}\left[\left(\frac{3\beta}{8} + D\beta G_1\right)q_1 + \left(\frac{\theta}{4} + D\beta G_2\right)p_1\right],$$

$$B_{20} = e^{-\beta\tau/2}\left[\left(\frac{1}{2}\langle p_c(0)\rangle\beta + \langle z_c(0)\rangle\right)q_1 + \langle p_c(0)\rangle\theta p_1\right]. \qquad (6.48)$$

Performing the inverse Fourier transform and taking the modulus we obtain,

$$|\rho_{1/2,-1/2}(R, r, \tau)| = \frac{|ab^*|}{\sqrt{\pi}\sigma_*}\, e^{\xi\eta^2}\, e^{-(r + r_0\eta)^2/2\tilde\sigma^2}\, e^{-(B_{20} - R)^2/4\sigma_*^2},$$

$$|\rho_{-1/2,1/2}(R, r, \tau)| = \frac{|a^*b|}{\sqrt{\pi}\sigma_*}\, e^{\xi\eta^2}\, e^{-(r - r_0\eta)^2/2\tilde\sigma^2}\, e^{-(B_{20} - R)^2/4\sigma_*^2},$$

$$(6.49)$$

where

$$\tilde{\sigma}^2 = \frac{2\sigma_*^2}{4\sigma_*^2 C_{12} - B_{11}^2},$$

$$r_0 = \frac{2\sigma_*^2 C_{11} - B_{11} B_{10}}{4\sigma_*^2 C_{12} - B_{11}^2}, \tag{6.50}$$

$$\xi = \frac{B_{10}^2}{4\sigma_*^2} - C_{10} - \frac{r_0^2}{2\tilde{\sigma}^2}.$$

The maxima are located at

$$(R = B_{20}, \ r = -\eta r_0) \qquad \text{for} \qquad |\rho_{1/2,-1/2}|$$

and at

$$(R = B_{20}, \ r = \eta r_0) \qquad \text{for} \qquad |\rho_{-1/2,1/2}|.$$

In (z_c, z_c') coordinates this corresponds to

$$M_{+-} = (z_c = B_{20} - \eta r_0/2, \ z_c' = B_{20} + \eta r_0/2) \qquad \text{for} \qquad |\rho_{1/2,-1/2}|$$

and

$$M_{-+} = (z_c = B_{20} + \eta r_0/2, \ z_c' = B_{20} - \eta r_0/2) \qquad \text{for} \qquad |\rho_{-1/2,+1/2}|,$$

so that the distance between them is given by $\Delta_{nd} = \sqrt{2}\eta|r_0|$. Next, we consider the quadratic form $(r \pm r_0\eta)/2\tilde{\sigma}^2 + (B_{20} - R)^2/4\sigma_*^2$ in the $z_c - z_c'$ plane. Straightforward calculations show that this is an ellipse whose semi-axes are respectively given by $\tilde{\sigma}$ (across the diagonal) and $2\sqrt{2}\sigma_*$ (along the diagonal). The centers of the peaks M_{+-}, M_{-+} are symmetric with respect to the diagonal line $z_c = z_c'$.

The most remarkable difference compared with the diagonal case, is the presence of irreversible decoherence. Indeed, the heights of the peaks are exponentially reduced in time by the damping factor $\sim \exp(-4\eta^2 D\beta\tau)$. This, in turn, defines a characteristic time scale of decoherence: $\tau_d = 1/4\eta^2 D\beta$.

This formula exactly coincides with the expression (5.9) based on a semi-qualitative analysis. The value of the decoherence time $T_d = \tau_d/\omega_c$ is very small: $T_d \approx 60$ ps at the temperature $T = 1$ mK.

At time $\tau = \tau_0$, when two diagonal peaks are clearly separated, the damping factor is $4\eta^2 D\beta\tau_0$. We expect to observe the coherence between the two peaks if this factor is not much more than one unit. Thus, using the expression for τ_0 from Eq. (6.39), we can estimate the condition for the quantum coherence as $D\beta\eta^{3/2} < 1$, or

$$T < T'_{max} = \frac{Q}{k_B} \frac{(\hbar\omega_c)^{7/4} k_c^{3/4}}{F^{3/2}}. \tag{6.51}$$

For our parameters the value of T'_{max} is approximately 3×10^{-7} K.

Now, we will check the validity of our estimate. Equation (6.39) is valid if $\eta >> (D\beta)^2/\sqrt{8}$. Setting $T = T'_{max}$ or $D\beta\eta^{3/2} = 1$, we obtain $\eta^4 >> 1/\sqrt{8}$, which is definitely true, assuming $\eta >> 1$. Next, the condition of the validity of the high temperature approximation is $D >> 1$. For $T = T'_{max}$, it follows that $\eta^{3/2} << Q$. This inequality is roughly satisfied ($\eta^{3/2} = 1700$, $Q = 6700$). Finally, as we mentioned previously, the master equation fails at times $t \leq \hbar/k_B T$. Thus, the time considered, $\tau_0 = 2^{1/4}/\sqrt{\eta}$, must be much greater than $1/D$, which is definitely wrong. Thus, our condition (6.51) for the creation of the macroscopic quantum superposition (the Schrödinger cat state) is not justified for the parameters considered.

However, the tenfold increase of Q ($Q = 67000$) would increase the maximum temperature T'_{max} to 3 μK. At this temperature we obtain $\tau_0 \approx 0.1$ and $1/D \approx 0.027$. Thus, in this case, $\tau_0 \gg 1/D$, and all conditions of applicability of our equations are satisfied.

Chapter 7

Periodic Spin Reversals in Magnetic Resonance Force Microscopy (MRFM) Driven by π-Pulses

Now we transfer to the main topic of our book — the theory of MRFM. In this chapter we will consider a simple (from the theoretical point of view) MRFM technique: application of a sequence of resonant π-pulses, which drive the periodic reversals of the spin. In turn, the spin periodic reversals drive the cantilever vibrations, which are to be detected (Berman and Tsifrinovich [22]).

We will consider the same setup as in Chaps. 5 and 6 (see Fig. 5.1). In addition to the external magnetic field $\vec{B}_{ext}$ we assume a transversal *rf* rotating field $\vec{B}_1$, (see formulas (2.13)). Let assume that the spin magnetic moment $\vec{\mu}$ points initially in the positive z-direction. The total magnetic field on the spin contains three parts. The first one is the permanent external field $\vec{B}_{ext}$. The second one is the dipole field $\vec{B}_d$ produced by the CT, i.e. by the ferromagnetic particle on the CT. The third one is the *rf* field $\vec{B}_1$. The dipole field can be represented as a sum of two terms

$$\vec{B}_d = \vec{B}_d^{(0)} + \vec{B}_d^{(1)}, \tag{7.1}$$

where $\vec{B}_d^{(0)}$ corresponds to the equilibrium position of CT with no spin, and $\vec{B}_d^{(1)}$ is associated with the driven CT vibrations caused by the spin. We assume here that $\vec{B}_d^{(1)}$ is small compared to $\vec{B}_d^{(0)}$ and $\vec{B}_1$. We also choose the value of the *rf* frequency ω:

$$\omega = \gamma \left(B_{ext} + B_d^{(0)} \right). \tag{7.2}$$

It means that the *rf* field is resonant to the spin if we ignore the small contribution of $\vec{B}_d^{(1)}$. Let us transfer to the RSC, which rotates about the z-axis with frequency ω. As we have shown in Chap. 1 the effective magnetic field $\vec{B}_{eff}$ in the RSC is the permanent field of magnitude B_1, which lies in the transversal plane. The spin magnetic moment $\vec{\mu}$ will precess about the effective field $\vec{B}_{eff}$ with the Rabi frequency $\omega_R = \gamma B_1$. Let one applies, instead of the continuous *rf* field, a *rf* pulse of duration π/ω_R. Such a pulse is called the π-pulse: it reverses the direction of the magnetic moment $\vec{\mu}$ (see Fig. 7.1). If one applies a periodic sequence of π-pulses then the direction of the magnetic moment changes periodically with the period equal to the double time interval between the π-pulses. When the magnetic moment $\vec{\mu}$ points in the positive z-direction it attracts the ferromagnetic particle. If $\vec{\mu}$ points in the negative z-direction it repels the ferromagnetic particle. Thus, the periodic sequence of π-pulses generates the periodic magnetic force on the CT. If the period of this force matches the cantilever period $T_c = 2\pi/\omega_c$, then the resonant magnetic force will drive the resonant CT vibrations. If the amplitude of the driven CT vibrations exceeds the amplitude of the thermal CT vibrations it is possible to detect experimentally a single spin.

Below we estimate the amplitudes of the driven and thermal vibrations of the CT. We assume that the cantilever frequency ω_c is much smaller than the Rabi frequency ω_R. In this case the duration of the π-pulse π/ω_R is small compared to the time interval between the pulses $T_c/2$. Thus, we can approximate the z-component of the magnetic force $F_z(t)$ on the CT with a periodic rectangular function of amplitude F.

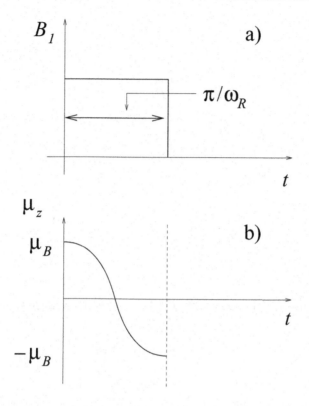

Figure 7.1: Action of the *rf* π-pulse: (a) amplitude of the *rf* field. (b) Change of μ_z.

Let us choose, for example, the even function (see Fig. 7.2):

$$F_z(t) = \begin{cases} F \text{ for } -\dfrac{T_c}{4} < t < \dfrac{T_c}{4}, \\[2ex] -F \text{ for } -\dfrac{T_c}{2} < t < -\dfrac{T_c}{4}, \text{ or } \dfrac{T_c}{4} < t < \dfrac{T_c}{2}. \end{cases} \tag{7.3}$$

The Fourier component of $F_z(t)$ on the CT frequency, ω_c, is $F_\omega \cos \omega_c t$, where F_ω is

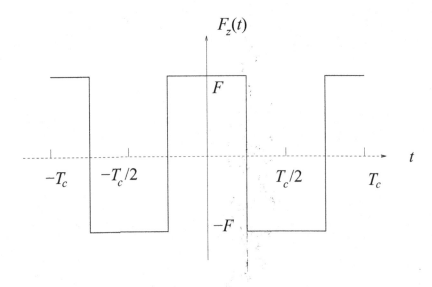

Figure 7.2: Periodic force $F_z(t)$ generated by a single spin.

$$F_\omega = \frac{2}{T_c} \int_{-T_c/2}^{T_c/2} F(t) \cos \omega_c t \, dt = \frac{4}{\pi} F. \qquad (7.4)$$

This component drives the resonant vibrations of the CT.

For the harmonic magnetic force $F_\omega \exp(i\omega t)$, the CT displacement is given by:

$$z_c = \frac{F_\omega/m^*}{\omega_c^2 - \omega^2 + i\omega^2/Q} \, e^{i\omega t}. \qquad (7.5)$$

If we take into considerations only the resonant Fourier component of the magnetic force then, putting $\omega = \omega_c$, we obtain:

$$z_c = -iQ(F_\omega/k_c) \, e^{i\omega t}. \qquad (7.6)$$

To estimate the amplitude of the driven CT vibrations we will take the magnetic field gradient $\partial B_d/\partial z = 100$ kT/m. Then the magnetic force:

$$F = \mu_B \frac{\partial B_d}{\partial z} = 940 \ z\text{N}. \qquad (7.7)$$

Further we will use the parameters from Ref. [17], as we did in the previous two chapters. Taking $Q = 6700$, $k_c = 6.5$ μN/m we obtain the amplitude A of the driven vibrations:

$$A = \frac{4}{\pi} \frac{QF}{k_c} = 1 \text{ nm}. \tag{7.8}$$

Next, we will estimate the root-mean-square (rms) amplitude, z_{rms} of the thermal CT vibrations. We will use the fluctuation-dissipation theorem (see, for example, the textbook of Landau and Lifshitz [23]). According to this theorem the value of z_{rms} is given by the expression

$$z_{rms} = \left[2\hbar Im(\chi) \coth\left(\frac{\hbar\omega_c}{2k_B T} \right) \Delta f \right]^{1/2}, \tag{7.9}$$

$$\Delta f = \frac{\omega_b}{2\pi},$$

where χ is the resonant susceptibility of the CT and ω_b is the bandwidth of the measuring device. It follows from (7.6) that the resonant susceptibility is

$$\chi = i\frac{Q}{k_c}. \tag{7.10}$$

Assuming $\hbar\omega_c \ll k_B T$ we simplify the formula (7.9):

$$z_{rms} \simeq \left(\frac{4k_B T Q \Delta f}{k_c \omega_c} \right)^{1/2}. \tag{7.11}$$

Note that putting $\Delta f = \omega_c/4Q$ we will obtain the estimate $z_{rms} = (k_B T/k_c)^{1/2}$ which follows from the equipartition theorem. Using the value $\Delta f = 0.4$ Hz from Ref. [17] we obtain the value $z_{rms} \approx 1.5\sqrt{T}$(nm), where the temperature T is taken in kelvins. The rms amplitude of the thermal vibrations is to be compared with the rms amplitude of the driven vibrations $A_{rms} \approx 0.71$ nm. It follows that for our parameters the single spin detection is possible for $T < 0.2$K.

The same result can be obtained if we compare the effective thermal force acting on the oscillator with the resonant component of the magnetic force

F_ω. According to the fluctuation-dissipation theorem the rms force is given by:

$$F_{rms} = \left\{ \frac{2\hbar Im(\chi)}{|\chi|^2} \coth\left(\frac{\hbar\omega_c}{2k_B T}\right) \Delta f \right\}^{1/2}. \tag{7.12}$$

Taking $\hbar\omega_c \ll k_B T$ we obtain

$$F_{rms} \approx \left(\frac{4k_B T k_c \Delta f}{Q\omega_c}\right)^{1/2} = \left(\frac{k_c}{Q}\right) z_{rms}. \tag{7.13}$$

Since the resonant component of the magnetic force is

$$F_\omega = \left(\frac{k_c}{Q}\right) A, \tag{7.14}$$

comparing $F_\omega/\sqrt{2}$ with F_{rms} will result in the same condition of the single-spin detection $T < 0.2$ K.

Note that we ignored the direct interaction between the spin and its environment, which will cause the quantum jumps — random change of the spin direction. Clearly, quantum jumps will prevent the observation of the driven CT vibrations if the characteristic time interval between two consecutive jumps is smaller than the cantilever time constant $T_r = Q/\omega_c$.

Chapter 8

Oscillating Adiabatic Spin Reversals Driven by the Frequency Modulated rf Field

In this chapter we consider a technique, which has been widely used in MRFM experiments (see, for example, Rugar *et al.* [6]). This technique is based on the frequency modulation of the rf field. To explain the idea we will consider the same setup as in the previous chapters (see Fig. 5.1) with an additional rotating transversal rf field. As we mentioned in Chap. 2 the frequency modulation of the rf field causes the change in the direction of the effective field in the RSC. If the frequency ω of the rf field matches the frequency ω_L of the Larmor precession of the spin ($\omega = \omega_L$) the z-component of the effective field is zero, and the effective field in the RSC equals the rf field: $\vec{B}_{eff} = \vec{B}_1$. If the frequency ω deviates from the value ω_L, the effective magnetic field acquires the z-component $(\omega_L - \omega)/\gamma$. Now, let the value of ω change periodically from $\omega_L + \Delta$ to $\omega_L - \Delta$, and $\Delta \gg \omega_R = \gamma B_1$. Then, the effective magnetic field experiences the cyclic reversals, (see Fig. 2.7). If the condition (2.27) for the adiabatic reversals is satisfied, then the spin component along the effective magnetic field is an approximate integral of motion (adiabatic invariant). It means that the spin being initially directed along the effective magnetic field will experience adiabatic reversals together with the effective field. If the period of the spin adiabatic reversals matches the CT period $T_c = 2\pi/\omega_c$, then the periodic magnetic force produced by the spin on the CT forces the driven cantilever vibrations in the same way as in the case of action of the periodic sequence of π-pulses described in the previous chapter. Note, that our reasoning is valid if the oscillating dipole field $B_d^{(1)}$, associated with the CT vibrations, is small compared to B_1, so that we can ignore the oscillations of the dipole field. However, in our theoretical analysis we take into consideration the oscillations of the dipole field. Below, in Sec. 8.1, we consider the entire CT-spin system within the framework of the Schrödinger equation (Berman et al. [24]). In Sec. 8.2, we discuss the decoherence and the thermal diffusion caused by the interaction between the cantilever and its environment (Berman et al. [25]).

8.1 Schrödinger dynamics of the CT-spin system

The transversal frequency modulated *rf* field acting on the spin system is represented as

$$
\begin{aligned}
B_{1x} &= B_1 \cos(\omega t + \phi(t)), \\
B_{1y} &= B_1 \sin(\omega t + \phi(t)),
\end{aligned}
\tag{8.1}
$$

where $\phi(t)$ describes a smooth change in phase required for cyclic adiabatic reversals of the spins: $|d\phi/dt| \ll \omega$.

The Hamiltonian of the spin-oscillator system in the RSC has the form:

$$
\mathcal{H} = \frac{1}{2m^*}\hat{p}_c^2 + \frac{1}{2}k_c z_c^2 + \hbar\left(\omega_L - \omega - \frac{d\phi}{dt}\right)\hat{S}_z + \hbar\omega_R\hat{S}_x - \gamma\hbar\frac{\partial B_d}{\partial z}\hat{S}_z z_c, \tag{8.2}
$$

where

$$
\omega_L = \gamma B_0, \qquad \omega_R = \gamma B_1. \tag{8.3}
$$

If we put $B_1 = 0$ and $\omega = d\phi/dt = 0$, the Hamiltonian (8.2) will transfer to the MFM Hamiltonian (6.2).

We will rewrite Hamiltonian (8.2) in the dimensionless form using the natural quantum units for energy, coordinate and momentum, introduced in Chap. 6:

$$\mathcal{H} = \frac{1}{2}\left(\hat{p}_c^2 + z_c^2\right) - \dot{\phi}\hat{S}_z + \epsilon\hat{S}_x - 2\eta\hat{S}_z z_c, \tag{8.4}$$

where

$$\dot{\phi} = d\phi/d\tau, \qquad \tau = \omega_c t, \qquad \epsilon = \omega_R/\omega_c, \tag{8.5}$$

and the parameter of the spin-CT interaction η is defined in (6.4).

The dimensionless Schrödinger equation can be written in the form,

$$\dot{\Psi} = \mathcal{H}\Psi, \tag{8.6}$$

where,

$$\Psi = \begin{pmatrix} \psi_1(z_c, \tau) \\ \psi_2(z_c, \tau) \end{pmatrix}, \tag{8.7}$$

is a dimensionless spinor, and $\dot{\Psi} = \partial\Psi/\partial\tau$. Next, we expand the functions, $\psi_1(z_c, \tau)$ and $\psi_2(z_c, \tau)$, in terms of the eigenfunctions, u_n, of the unperturbed oscillator Hamiltonian, $(\hat{p}_c^2 + z_c^2)/2$,

$$\psi_1(z_c, \tau) = \sum_{n=0}^{\infty} A_n(\tau)u_n(z_c), \qquad \psi_2(z_c, \tau) = \sum_{n=0}^{\infty} B_n(\tau)u_n(z_c),$$

$$u_n(z_c) = \pi^{1/4}2^{n/2}(n!)^{1/2}e^{-z_c^2/2}H_n(z_c), \tag{8.8}$$

where $H_n(z_c)$ is the Hermite polynomial. Substituting (8.8) in (8.6), we derive the coupled system of equations for the complex amplitudes, $A_n(\tau)$, and $B_n(\tau)$,

$$i\dot{A}_n = (n + \frac{1}{2} + \frac{\dot{\phi}}{2})A_n - \frac{\eta}{\sqrt{2}}(\sqrt{n}A_{n-1} + \sqrt{n+1}A_{n+1}) - \frac{\epsilon}{2}B_n,$$

$$i\dot{B}_n = (n + \frac{1}{2} + \frac{\dot{\phi}}{2})B_n + \frac{\eta}{\sqrt{2}}(\sqrt{n}B_{n-1} + \sqrt{n+1}B_{n+1}) - \frac{\epsilon}{2}A_n. \tag{8.9}$$

To derive Eqs. (8.9), we used (4.6), (4.7) and the following expressions for creation and annihilation operators,

$$\frac{1}{2}(\hat{p}_c^2 + z_c^2) = \hat{a}^\dagger \hat{a} + \frac{1}{2},$$

$$z_c = \frac{1}{\sqrt{2}}(\hat{a}^\dagger + \hat{a}), \qquad \hat{p}_c = \frac{i}{\sqrt{2}}(\hat{a}^\dagger - \hat{a}), \qquad (8.10)$$

$$[\hat{a}, \hat{a}^\dagger] = 1.$$

Below we describe the results of numerical simulations of the spin-oscillator dynamics, for the value of the spin-CT interaction parameter $\eta = 0.3$.

The initial state of the CT was chosen as a coherent state $|\alpha\rangle$, see Chap. 4, in the quasiclassical region of parameters $|\alpha| \gg 1$. Using formula (4.13), the initial conditions can be represented as

$$\psi_1(z_c, 0) = \sum_{n=0}^{\infty} A_n(0) u_n(z_c),$$

$$\psi_2(z_c, 0) = 0, \qquad (8.11)$$

$$A_n(0) = (\alpha^n/\sqrt{n!}) \exp\left(-|\alpha|^2/2\right).$$

The initial averages $\langle z_c(0)\rangle$ and $\langle p_c(0)\rangle$ can be expressed in terms of α using (4.9):

$$\langle z_c(0)\rangle = \frac{1}{\sqrt{2}}(\alpha^* + \alpha), \quad \langle p_c(0)\rangle = \frac{i}{\sqrt{2}}(\alpha^* - \alpha). \qquad (8.12)$$

The value of α was cosen to be $\alpha = -\sqrt{2} \cdot 10$, which corresponds to the initial average value $\langle n \rangle = |\alpha|^2 = 200$.

Note that the values of $|\alpha|$ cannot be significantly reduced if one simulates a quasiclassical cantilever. At the same time, increasing $|\alpha|$ one increases the number of states, n, involved in the dynamics which makes the simulations of quantum dynamics more complicated. The system of Eqs. (8.9) was integrated numerically using a standard Runge-Kutta fourth order method. The stability of the results has been checked by increasing the dimension of the

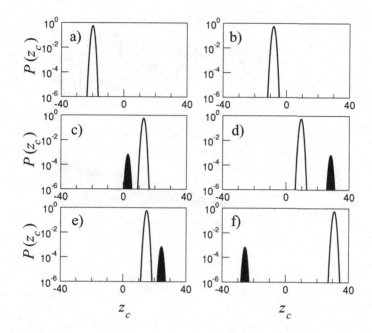

Figure 8.1: Probability distribution of the CT coordinate , z_c, for $\epsilon = 400$ and $\eta = 0.3$. The initial conditions $\langle z_c(0) \rangle = -20$, $\langle p_c(0) \rangle = 0$, $(\alpha = -\sqrt{2} \times 10)$, and the average spin is in the direction of the effective field. Times for (a,b,c,d,e,f) are respectively $\tau = 0, 20, 64.8, 104, 160, 221.6$.

oscillator basis (up to 3000 levels) and by decreasing the time integration step. Figure 8.1 shows the typical probability distribution

$$P(z_c, \tau) = |\psi_1(z_c, \tau)|^2 + |\psi_2(z_c, \tau)|^2, \qquad (8.13)$$

obtained from the numerical simulation of Eqs. (8.9) for six different instants of time, τ, and for the parameters $\eta = 0.3$ and $\epsilon = 400$ (the dimensionless period $\tau_c = 2\pi$ corresponds to the dimensional period $T_c = 2\pi/\omega_c$). This figure reveals that the cantilever can be found in two different positions. Indeed, near $\tau = 80$, the probability distribution (8.13) splits into two asymmetric peaks. After this, the separation between the peaks varies periodically in time. The ratio of the peak amplitudes is about 1000 for chosen parameters.

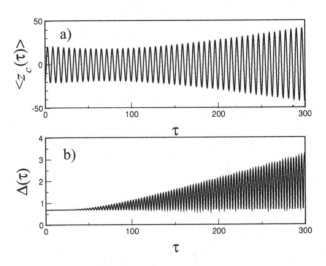

Figure 8.2: CT dynamics. (a) Average CT coordinate as a function of τ and (b) its standard deviation $\Delta(\tau) = [\langle z_c^2(\tau) \rangle - \langle z_c(\tau) \rangle^2]^{1/2}$. Parameters are the same as in Fig. 8.1.

(Hence, the amplitudes are shown in the logarithmic scale.)

The cyclic adiabatic inversion parameters were chosen,

$$\dot{\phi} = \begin{cases} -6000 + 300\tau, & \text{if } \tau \leq 20, \\ A\sin(\tau - 20), & \text{if } \tau > 20, \end{cases} \qquad (8.14)$$

where $A = 1000$, so that the standard condition for the adiabatic motion (2.27), which in our parameters can be written as $|\ddot{\phi}| \ll \epsilon^2$, is satisfied. The chosen parameters in Eq. (8.14) allow one to "catch" the spin, initially oriented in the positive (or negative) z-direction, by the effective magnetic field, and to put it approximately in the positive (or negative) x-direction at $\tau = 20$. For times $\tau > 20$, the spin oscillates in the $x - z$-plane, together with the effective magnetic field. It is clear that the small peak does not significantly influence the average CT coordinate. Figure 8.2 shows the average CT coordinate, $\langle z_c(\tau) \rangle$, and the corresponding standard deviation, $\Delta(\tau) = [\langle z_c^2 \rangle - \langle z_c \rangle^2]^{1/2}$. One can see a fast increase of the average amplitude of the CT vibrations, while the standard deviation still remains small. This, in fact, is related to the initial conditions of the spin, which was taken in the direction of the z-axis. For instance, if the spin initially points in the x-axis ($\psi_1(z_c, 0) = \psi_2(z_c, 0)$), the calculations show two large peaks with equal amplitudes.

The two peaks in the probability distribution, shown in Fig. 8.1 indicate two possible trajectories of the cantilever (similar to the Stern-Gerlach effect). The two peaks are well-separated for shown instants of time. When

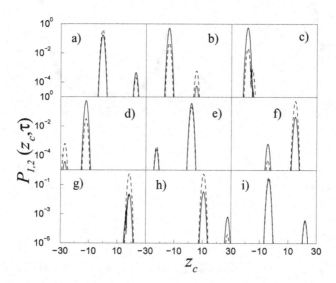

Figure 8.3: Probability distributions, $P_1(z_c, \tau) = |\psi_1(z_c, \tau)|^2$ (solid curves), and $P_2(z_c, \tau) = |\psi_2(z_c, \tau)|^2$ (dashed curves) for nine instants of time: $\tau_k = 92.08 + 0.8k$, $k = 0, 1, ..., 8$.

the probability distribution splits into these peaks, the distance, d, between them initially increases. Then, d decreases so that the two peaks eventually overlap. After this, the probability distribution splits again so that the position of the minor peak is on the opposite side of the major peak. Again, the distance, d, first increases, then decreases until the two peaks overlap. This cycle repeats for as long as the simulations are run.

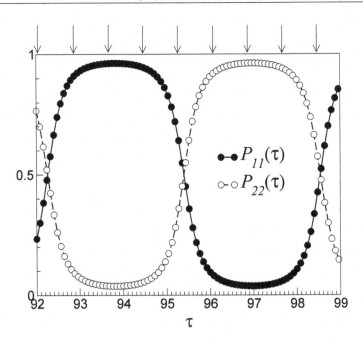

Figure 8.4: Integrated probability distributions of the spin z-components: $P_{11}(\tau)$, for $S_z = 1/2$ ($\bullet$); and $P_{22}(\tau)$, for $S_z = -1/2$ ($\circ$), as functions of time. Vertical arrows show the time instants, $\tau_k = 92.08 + 0.8k$, $k = 0, 1, ..., 8$ depicted in Fig. 8.3.

One might expect that the two peaks are associated with the functions $P_n(z_c, \tau) = |\psi_n(z_c, \tau)|^2$, $n = 1, 2$. In fact, the situation is more subtle: each function, $P_n(z_c, \tau)$ splits into two peaks. Figure 8.3, shows these two functions for nine instants of time: $\tau_k = 92.08 + 0.8k$, $k = 0, 1, ..., 8$ during one period of the cantilever vibration. One can see the splitting of both $P_1(z_c, \tau)$ and $P_2(z_c, \tau)$; the two peaks of the function $P_1(z_c, \tau)$ have the same positions as the two peaks of $P_2(z_c, \tau)$, but the amplitudes of these peaks differ. For instance, for $k = 1$ ($\tau = 92.88$) the left-hand peak is dominantly composed of $P_1(z_c, \tau)$, while the right hand peak is mainly composed of $P_2(z_c, \tau)$. Figure 8.4 shows the spatially integrated probability distributions: $P_{11}(\tau) = \int P_1(z_c, \tau) dz_c$ and $P_{22}(\tau) = \int P_2(z_c, \tau) dz_c$, as "truly continuous"

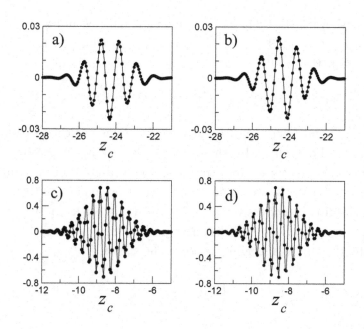

Figure 8.5: "Wave functions" belonging to "big" and "small" peaks. (a) circles: $Re(-\kappa\psi_1^{sm})$, solid line: $Re(\psi_2^{sm})$; (b) circles: $Im(-\kappa\psi_1^{sm})$, solid line: $Im(\psi_2^{sm})$; (c) circles: $Re(\kappa\psi_2^{b})$, solid line: $Re(\psi_1^{b})$; (d) circles: $Im(\kappa\psi_2^{b})$, solid: $Im(\psi_1^{b})$, where $\kappa(\tau = 76) = -2.9$.

functions of time, τ. (Vertical arrows show the time instants, τ_k.) The crucial problem is the following: Do the two peaks of the CT distribution correspond to the definite spin states? To answer this question the structure of the wave function of the CT-spin system has been studied. As was already mentioned, both functions, $\psi_1(z_c, \tau)$ and $\psi_2(z_c, \tau)$, contribute to each peak (see Fig. 8.3). When the two peaks are clearly separated one can represent each of these functions as a sum of two terms, corresponding to the "big" and "small" peaks,

$$\psi_{1,2}(z_c, \tau) = \psi_{1,2}^{b}(z_c, \tau) + \psi_{1,2}^{sm}(z_c, \tau). \tag{8.15}$$

It was found that with accuracy up to 1% the ratio, $\psi_2^{sm}(z_c, \tau)/\psi_1^{sm}(z_c, \tau) = -\psi_1^{b}(z_c, \tau)/\psi_2^{b}(z_c, \tau) = \kappa(\tau)$, where $\kappa(\tau)$ is a real function independent of z_c.

Results are shown in Fig. 8.5, for the same parameters as in Fig. 8.1, and for $\tau = 76$ with $\kappa(\tau) = -2.9$ obtained by a best fit procedure.

As a result, the total wave function can be represented in the form,

$$\Psi(z_c, s, \tau) = \psi^b(z_c, \tau)\chi^b(s, \tau) + \psi^{sm}(z_c, \tau)\chi^{sm}(s, \tau), \qquad (8.16)$$

where $\chi^b(s, \tau)$ and $\chi^{sm}(s, \tau)$ are spin wave functions, which are orthogonal to each other. Equation (8.16) shows that each peak in the probability distribution of the CT coordinate corresponds to a definite spin wave function. It was found that the average spin, $\langle \chi^b | \vec{S} | \chi^b \rangle$, corresponding to the big peak points in the direction of the vector $(\epsilon, 0, -d\phi/d\tau)$, whereas the average spin, $\langle \chi^{sm} | \vec{S} | \chi^{sm} \rangle$, corresponding to the small peak, points in the opposite direction. Note, that up to a small term, $2\eta z_c$, the vector,

$$\left(\epsilon, 0, -\frac{d\phi}{d\tau} \right),$$

is the effective field acting on the spin in units of ω_c/γ.

Figure 8.6, demonstrates the directions of the effective field (thick solid arrow); the direction of the average spin calculated using the $\chi^b(s, \tau)$ spin function (thin arrow); and the direction of the average spin calculated using the $\chi^{sm}(s, \tau)$ spin wave function (thin dashed arrow). This can only be done when the probability distributions corresponding to the small and big peaks, ψ^b and ψ^{sm}, are well separated in space. This is not the case in Figs. 8.3(c) and 8.3(g). Figures 8.6(c) and (g) represent the total average spin only (as a thin line). One should also take into account that the "lengths" of the effective magnetic field and the average spin of the small head has been renormalized respectively to the length 1 and $1/2$, in order to be put on the same scale (they are respectively a few orders of magnitude larger and smaller than the total average spin). The results presented in Fig. 8.6 allow one a better understanding of the structure of the total wave function described by Eq. (8.16).

The ratio of the integrated probabilities, $\int P(z_c, \tau)dz_c$, for the small and big peaks ($\sim 10^{-3}$ in Fig. 8.1) can be easily estimated as $\tan^2(\Theta/2)$, where

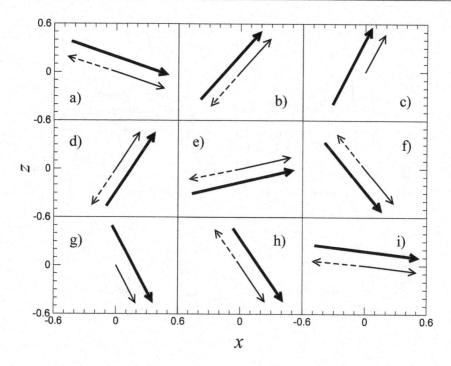

Figure 8.6: The directions of the effective magnetic field (thick solid arrow) renormalized to the unit length; the direction of the average spin calculated using the $\chi^b(s,\tau)$ spin function (thin arrow); and the direction of the average spin calculated using the $\chi^{sm}(s,\tau)$ spin function (thin dashed arrow) renormalized to the length $1/2$, in order to be plotted in the same picture. Times and parameters are as in Fig. 8.3. In Figs. (c) and (g) one single thin line has been drawn, for the total average spin. This is due to the spatial overlapping of the probability distributions corresponding to the small and big peak.

$\Theta \approx 0.07$ is the initial angle between the effective field, $(\epsilon, 0, -d\phi/d\tau)$, and the spin direction. Therefore, by measuring the cantilever vibrations, one finds the spin in a definite state in- or opposite to- the effective field. Numerical simulations for such a new initial condition, i.e. when the average spin points in- or opposite to- the effective field, are shown in Fig. 8.7. The probability

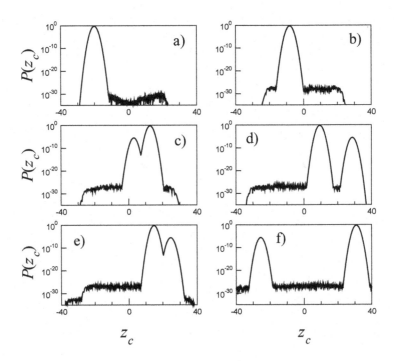

Figure 8.7: Probability distribution of the CT coordinate, z_c, for $\epsilon = 400$ and $\eta = 0.3$. The initial conditions: $\langle z_c(0) \rangle = -20$, $\langle p_c(0) \rangle = 0$ ($\alpha = -\sqrt{2} \times 10$), and the average spin is in the direction of the effective field.

distribution $P(z_c, \tau)$ shows again two peaks but the ratio of the integrated probabilities of these peaks is much less than in Fig. 8.1. ($\sim 10^{-6}$).

Note that Fig. 8.7 has a larger scale on the y-axis than that in Fig. 8.1, in order to show that the small peak is clearly beyond the unavoidable numerical errors (below 10^{-25} in Fig. 8.7).

Thus, for chosen parameters, the probability of the second peak in the CT position generated by a single spin measurement is small. This implies that the appearance of this peak cannot prevent the amplification of the cantilever vibration amplitude, and therefore the measurement of the state of a single spin.

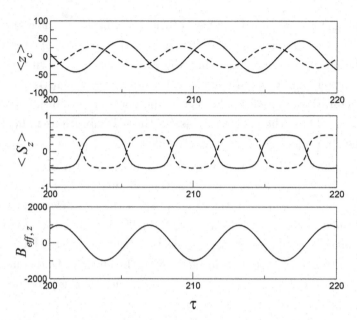

Figure 8.8: Measurement of the single-spin state using the phase of the CT vibrations. For the dynamics of $\langle z_c(\tau) \rangle$ and $\langle S_z(\tau) \rangle$ the solid line corresponds to the "big" peak of the probability distribution, and the dashed line corresponds to the "small" peak, renormalized to the similar amplitudes. At the bottom, the dynamics of the z-component of the effective field is shown.

So far, the described picture reminds the well-known Stern-Gerlach effect in which the cantilever measures the spin component not in the $z-$direction but along the effective magnetic field. An appearance of the second peak, even if the average spin points initially in the direction of the effective field, provides a difference with the Stern-Gerlach effect. The origin of this peak is a small deviations from the adiabatic motion of the spin even at a large amplitude of the effective field, and the back reaction of the CT vibrations on the spin.

The important question is the following: How to use the described technique not only to detect a spin signal but also to measure the state of a single spin? Let us consider the phase of the CT vibrations when the initial spin points in- or opposite to- the direction of the effective magnetic field. The computer simulations show that the phases of the CT vibrations for these two initial conditions are significantly different. When the amplitude of the CT vibrations increases, the phase difference for two initial conditions approaches π. Thus, the classical phase of the CT vibrations indicates the state of the spin relatively to the effective field.

Figure 8.8 demonstrates a process of measurement of a single-spin state using the phase of the CT vibrations. For the dynamics of $\langle z_c(\tau) \rangle$ and $\langle S_z(\tau) \rangle$ the solid curve corresponds to the "big" peak of the cantilever distribution, and the dashed curve corresponds to the "small" peak. At the bottom, the dynamics of the z-component of the effective field is shown. One can see that the solid curve of $\langle S_z(\tau) \rangle$ is in phase with the effective field component, $B_{eff,z}(\tau)$. The phase difference of the CT vibrations corresponding to two peaks approaches π for large times.

In practical applications it would be very desirable to use MRFM for measurement of the initial z-component of the spin. For this purpose one should provide the initial direction of the effective field to be the z-direction. Then, the initial z-component of the spin will coincide with its component relatively to the effective field. In computer simulations presented in Figs. 8.1–8.6 an instantaneous increase of the amplitude of the rf field, at $\tau = 0$ has been assumed. This causes an initial angle between the directions of the spin and the effective field, $\Theta \approx \epsilon/|d\phi/d\tau| \approx 0.07$. To eliminate this initial angle the quantum spin-cantilever dynamics for an adiabatic increase of the rf field amplitude has been simulated:

$$\epsilon = 20\tau \quad \text{for} \quad \tau \leq 20, \quad \text{and} \quad \epsilon = 400 \quad \text{for} \quad \tau > 20. \qquad (8.17)$$

Dependence for $d\phi/d\tau$ was taken the same as before. The results of these simulations are qualitatively similar to those presented in Figs. 8.1–8.6, but the integrated probability of the small peak was reduced to its residual value $\sim 10^{-6}$.

8.2 Decoherence and thermal diffusion for the CT

In order to consider the processes of decoherence and thermal diffusion for the CT we will use the master equation for the spin-CT system in the same way as it was done in Chap. 6. In the presence of the frequency modulated *rf* field the master equation (6.5) transforms into the equation

$$\frac{\partial \rho_{s,s'}(z_c, z_c', \tau)}{\partial \tau} = \left[\frac{i}{2}(\partial_{zz} - \partial_{z'z'}) - \frac{i}{2}(z_c^2 - z_c'^2) - \frac{\beta}{2}(z_c - z_c')(\partial_z - \partial_{z'}) \right.$$

$$\left. - D\beta(z_c - z_c')^2 - 2i\eta(z_c's' - z_c s) + i\dot{\phi}(s' - s) \right] \rho_{s,s'}(z_c, z_c', \tau) \qquad (8.18)$$

$$- i\frac{\epsilon}{2}\left[\rho_{s,-s'}(z_c, z_c', \tau) - \rho_{-s,s'}(z_c, z_c', \tau) \right].$$

For computer simulations it is convenient to expand the density matrix $\rho_{s,s'}(z_c, z_c', \tau)$ over the product of the eigenfunctions of the oscillator's Hamiltonian:

$$\rho_{s,s'}(z_c, z_c', \tau) = \sum_{n,m} A_{n,m}^{s,s'}(\tau)\, u_n(z_c) u_m^*(z_c'). \qquad (8.19)$$

Substituting this expansion into the master equation, we obtain the system of equations for the amplitudes $A_{n,m}^{s,s'}(\tau)$:

$$\dot{A}_{n,m}^{s,s'}(\tau) = \left[i\dot{\phi}(\tau)(s' - s) + \frac{1}{2}\beta - (n + m + 1)D\beta - i(n - m) \right] A_{n,m}^{s,s'}(\tau) -$$

$$i\eta s' \sqrt{2m} A_{n,m-1}^{s,s'}(\tau) - i\eta s' \sqrt{2m+2} A_{n,m+1}^{s,s'}(\tau) +$$

$$i\eta s \sqrt{2n} A_{n-1,m}^{s,s'}(\tau) + i\eta s \sqrt{2n+2} A_{n+1,m}^{s,s'}(\tau) +$$

$$D\beta \sqrt{m(n+1)} A_{n+1,m-1}^{s,s'}(\tau) + D\beta \sqrt{n(m+1)} A_{n-1,m+1}^{s,s'}(\tau) +$$

$$(D + \frac{1}{2})\beta\sqrt{(n+1)(m+1)}A^{s,s'}_{n+1,m+1}(\tau) + (D - \frac{1}{2})\beta\sqrt{nm}A^{s,s'}_{n-1,m-1}(\tau) -$$

$$\frac{\beta}{2}(D - \frac{1}{2})\sqrt{n(n-1)}A^{s,s'}_{n-2,m}(\tau) - \frac{\beta}{2}(D + \frac{1}{2})\sqrt{(n+1)(n+2)}A^{s,s'}_{n+2,m}(\tau) -$$

$$\frac{\beta}{2}(D - \frac{1}{2})\sqrt{m(m-1)}A^{s,s'}_{n,m-2}(\tau) -$$

$$\frac{\beta}{2}(D + \frac{1}{2})\sqrt{(m+2)(m+1)}A^{s,s'}_{n,m+2}(\tau) - \frac{i}{2}\epsilon[A^{s,-s'}_{n,m}(\tau) - A^{-s,s'}_{n,m}(\tau)]. \quad (8.20)$$

First, we describe the results of numerical solution of Eqs. (8.20) for $\beta = D = 0$ and the same values of parameters η and ϵ, as in Sec. 8.1. The initial conditions for the density matrix correspond to Eqs. 8.11:

$$\rho_{s,s'}(z_c, z'_c, 0) = \psi_1(z_c, 0)\psi_1^*(z'_c, 0)\hat{\chi}_{ss'}(0), \quad (8.21)$$

where $\hat{\chi}(0)$ is the spin density matrix:

$$\hat{\chi}(0) = \begin{pmatrix} 1 & 0 \\ 0 & 0 \end{pmatrix}, \quad (8.22)$$

which describes the spin pointing in the positive z−direction. The modulus of the parameter α for the initial coherent state (8.11) was taken five times smaller than in Sec. 8.1: $|\alpha| = 2\sqrt{2}$, instead of $|\alpha| = 10\sqrt{2}$, in order to deal with greater amount of computations.

For $\tau > 0$, the density matrix describes the entangled state which cannot be represented as a product of the cantilever and spin parts. The initial peak of $\rho_{s,s'}(z_c, z'_c, \tau)$ splits into two peaks which are centered along the diagonal $z_c = z'_c$, and two peaks centered at $z_c \neq z'_c$, off the diagonal. The density matrix can be represented approximately as a sum of four terms corresponding to four peaks,

$$\rho_{s,s'}(z_c, z'_c, \tau) = \rho^{(1)}_{s,s'} + \rho^{(2)}_{s,s'} + \rho^{(3)}_{s,s'} + \rho^{(4)}_{s,s'}, \quad (8.23)$$

where we omit variables, z_c, z'_c, τ. The matrices, $\rho^{(1)}$ and $\rho^{(2)}$, describe the "big" and "small" diagonal peaks; $\rho^{(3)}$ and $\rho^{(4)}$, describe the peaks centered at $z_c \neq z'_c$.

As an illustration, we show in Fig. 8.9, the quantity,

$$\ln |\rho_{1/2,1/2}(z_c, z'_c, \tau) + \rho_{-1/2,-1/2}(z_c, z'_c, \tau)|. \tag{8.24}$$

We have found that with accuracy to 1% the density matrix, $\rho^{(1)}_{s,s'}(z_c, z'_c, \tau)$, can be represented as a product of the coordinate and spin parts

$$\rho^{(1)}_{s,s'}(z_c, z'_c, \tau) = \hat{R}^{(1)}(z_c, z'_c, \tau)\hat{\chi}^{(1)}_{s,s'}(\tau), \tag{8.25}$$

where $\hat{\chi}^{(1)}_{s,s'}(\tau)$ describes the spin which points in the direction of the effective field $(\epsilon, 0, -\dot{\phi}(\tau))$. A similar expression is valid for $\rho^{(2)}_{s,s'}(z_c, z'_c, \tau)$. But in this case, $\hat{\chi}^{(2)}_{s,s'}(\tau)$ describes a spin which points in the opposite direction.

Next we describe the evolution of the density matrix for the finite values of the parameters β and D: $\beta = 10^{-3}$ and $D = 10$. (The high-temperature approximation for the master equation requires the inequality $D \gg 1$, which is satisfied for $D = 10$.) The initial uncertainty of the CT position is, $\delta z_c = 1/\sqrt{2}$. Due to thermal diffusion, the uncertainty of the CT position increases with time. Thus, we have two effects: i) the increase of the amplitude of the driven cantilever vibrations (similar to the Hamiltonian dynamics) and ii) the increase of the uncertainty of the CT position due to the thermal diffusion. If the second effect dominates, the two positions of the diagonal peaks (i.e. peaks centered on the line $z_c = z'_c$) become indistinguishable. In this case, one cannot provide a spin measurement with two possible outcomes.

It was found that peaks centered on the diagonal retain the main properties described by the Hamiltonian dynamics. The "density matrix",

$$\rho^{(k)}_{s,s'}(z_c, z'_c, \tau),$$

for $k = 1, 2$ can be approximately represented as a product of the CT and spin parts. The spin part of the matrix describes the spin which points in the direction of the external effective field (for $k = 1$) or in the opposite direction (for $k = 2$).

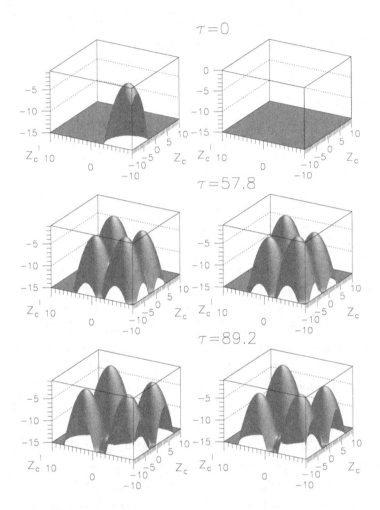

Figure 8.9: Left column: three-dimensional plot of $\ln|\rho_{1/2,1/2}(z_c, z'_c, \tau) + \rho_{-1/2,-1/2}(z_c, z'_c, \tau)|$, for different times τ. Right column: three-dimensional plot of $\ln|\rho_{1/2,-1/2}(z_c, z'_c, \tau) + \rho_{-1/2,1/2}(z_c, z'_c, \tau)|$, at the same times τ. The values of parameters are: $\epsilon = 400$, $\eta = 0.3$, $\beta = D = 0$. The initial conditions are: $\langle z_c(0) \rangle = -4$, $\langle p_c(0) \rangle = 0$.

Next, we discuss the two peaks centered at $z_c \neq z_c'$. As an illustration, Fig. 8.10 shows the quantities,

$$\ln |\rho_{1/2,1/2}(z_c, z_c', \tau) + \rho_{-1/2,-1/2}(z_c, z_c', \tau)|, \qquad (8.26)$$

and

$$\ln |\rho_{1/2,-1/2}(z_c, z_c', \tau) + \rho_{-1/2,1/2}(z_c, z_c', \tau)|, \qquad (8.27)$$

at $\tau = 57.8$. One can see the small peaks centered at $z_c \neq z_c'$. The peaks centered at $z_c \neq z_c'$ describe the coherence between the two CT positions. The amplitude of these peaks quickly decreases due to the decoherence. Thus, the master equation explicitly describes the process of measurement. The coherence between two cantilever trajectories (the macroscopic Schrödinger cat state) quickly disappears. As a result, the cantilever will "choose" one of two possible trajectories. Correspondingly, (depending on the cantilever trajectory) the spin will point in the direction of the effective magnetic field or in the opposite direction.

In conclusion, we note that the computer simulations of the CT-spin dynamics using the master equation allow one to understand the basic elements of the single-spin measurement process, which contains the driven vibrations of the CT, decoherence and thermal diffusion. From the other side, because of the huge amount of computations the simulations of the driven CT vibrations at $\tau > Q$ and for realistic values of parameters seems to be a very complicated problem. We also note that in this chapter as well as in the previous one we have ignored the direct interaction between the spin and its environment, which causes the quantum jumps of the spin (spin flips).

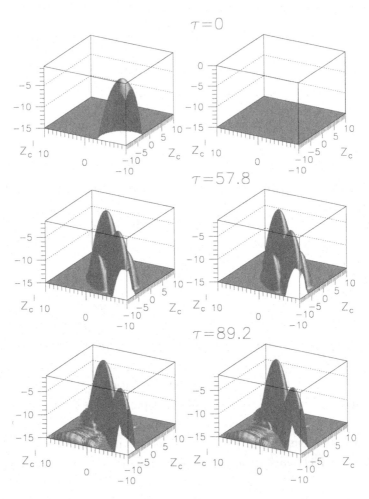

Figure 8.10: Left column: the three-dimensional plot for $\ln|\rho_{1/2,1/2}(z_c, z'_c, \tau) + \rho_{-1/2,-1/2}(z_c, z'_c, \tau)|$, at different times τ. Right column: the three-dimensional plot for $\ln|\rho_{1/2,-1/2}(z_c, z'_c, \tau) + \rho_{-1/2,1/2}(z_c, z'_c, \tau)|$, at different times τ. The values of parameters are: $\epsilon = 400$, $\eta = 0.3$, $\beta = 10^{-3}$, $D = 10$. Initial conditions are: $\langle z_c(0) \rangle = -4$, $\langle p_c(0) \rangle = 0$.

Chapter 9

Oscillating Cantilever-Driven Adiabatic Reversals (OSCAR) Technique in MRFM

This and the two next chapters are devoted to the brilliant technique, which has been used by Rugar *et al.* [8] for the first detection of a single electron spin below the surface of a non-transparent solid. This technique is called "the oscillating cantilever driven adiabatic reversals" (OSCAR). In the MRFM techniques considered in the previous two chapters the cantilever vibrations were driven by the periodic reversals of the spin. In turn, the periodic reversals of the spin were driven by the pulsed *rf* field (Chap. 7) or by the frequency modulated *rf* field (Chap. 8). In both cases, the oscillating dipole field $\vec{B}_d^{(1)}$ on the spin, associated with the CT vibrations, was small compared to the *rf* field, $\vec{B}_1$.

In the OSCAR technique the CT vibrations are driven by a feedback loop designed to keep the CT amplitude A constant. The amplitude of the oscillating dipole field $\vec{B}_d^{(1)}$ is much greater than the magnitude of the rotating *rf* field $\vec{B}_1$. As a result, the CT vibrations cause the cyclic adiabatic reversals of the spin with the CT frequency ω_c. In turn, the periodic reversals of the spin, produce the back resonant force on the CT. This force is, approximately

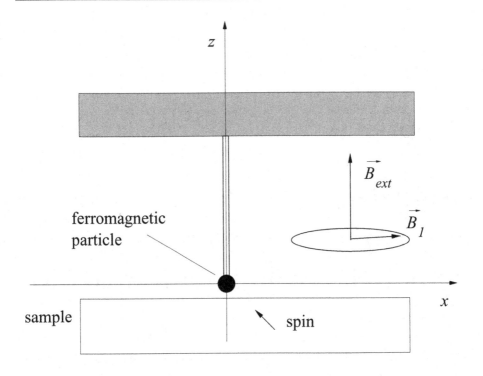

Figure 9.1: MRFM setup with the CT oscillating along the x-axis parallel to the sample surface.

proportional to the CT displacement and, consequently, causes a shift of the effective spring constant δk_c and the CT vibration frequency $\delta \omega_c$ The frequency shift $\delta \omega_c$ can be measured experimentally with a high precision.

The OSCAR technique has been introduced by Stipe *et al.* [26] in 2001. Two years later Mamin *et al.* [27] demonstrated the two-spin sensitivity in OSCAR. Finally, in 2004 Rugar *et al.* [8] reported the long-expected single-spin detection. In the OSCAR technique the CT may oscillate perpendicular or parallel to the sample surface. In the first case the cantilever itself is parallel to the sample surface, as it is shown in Fig. 5.1. In the second case the cantilever is placed perpendicular to the sample surface (see Fig. 9.1).

Below, in Sec. 9.1, we discuss the process of a single spin measurement using the OSCAR technique and estimate the main parameters of the CT-spin dynamics in the spirit of the mean field approximation (Berman *et al.* [28]). In Sec. 9.2, we describe shortly the experiment [8] on a single spin detection.

9.1 CT-spin dynamics: discussion and estimates

We will consider the CT oscillating parallel to the surface of the sample and the electron spin in the $x - z$ plane (Fig. 9.1). We assume that the electron spin is initially in its ground state, i.e. it points in the negative z-direction. It means that the external magnetic field $\vec{B}_{ext}$, which points in the positive z-direction, is much greater than the dipole field $\vec{B}_d$ produced by the ferromagnetic particle. Let us assume, for example, that the *rf* field is turned on when the CT is in its end position $\langle x_c \rangle = A$ and the z-component of the dipole field is greater than it was at $\langle x_c \rangle = 0$.

We may ignore the x-component of the dipole field if $B_{ext} \gg B_d$. Indeed, for the magnitude of the vector $\vec{B}_{ext} + \vec{B}_d$ we have:

$$\sqrt{(B_{ext} + B_{dz})^2 + B_{dx}^2} \approx B_{ext} + B_{dz}. \tag{9.1}$$

The y-component of the dipole field in Fig. 9.1 is zero since the spin is in the $x - z$ plane. If the z-component of the magnetic field $B_{ext} + B_{dz}$ for $\langle x_c \rangle > 0$ is greater than it was for $\langle x_c \rangle = 0$, the effective field in the RSC points initially, approximately, in the positive z-direction, and the spin $\langle \vec{S} \rangle$ is antiparallel to the effective field.

In the process of adiabatic motion, the spin $\langle \vec{S} \rangle$ remains antiparallel to the effective field. The z-component of the spin magnetic moment $\mu_z = -\gamma \hbar \langle S_z \rangle$ oscillates with the CT frequency. It produces a back resonant force on the CT. In Chap. 4 we have shown that this force is given by the expression (4.17), in which the gradient of the magnetic field is taken at the spin location when the CT is in the origin $\langle x_c \rangle = 0$. As the resonant back force is associated with the z-component of the magnetic moment μ_z, and the only component of the force, which affects the CT vibrations is the x-component F_x, we consider the approximate expression for F_x:

$$F_x \approx -\mu_z \frac{\partial B_z}{\partial x} \equiv -G_z \mu_z, \qquad (9.2)$$

where we introduced the notation $G_z = \partial B_z / \partial x < 0$.

For the estimation, below we will take, for example, parameters from Ref. [27]:

- The effective CT spring constant $k_c = 600 \ \mu\text{N/m}$.

- The CT frequency and period $\omega_c / 2\pi = 6.6$ kHz, $T_c = 150 \ \mu\text{s}$.

- The CT quality factor $Q = 5 \times 10^4$.

- The CT amplitude $A = 10$ nm.

- The rotating rf field amplitude and frequency $B_1 = 300 \ \mu\text{T}$, $\omega = 3$ GHz, $(\omega/\gamma = 100 \ \text{mT})$.

- The Rabi frequency and period $\omega_R / 2\pi = \gamma B_1 / 2\pi = 8.4$ MHz, $T_R = 2\pi / \omega_R = 120$ ns.

- The magnetic field gradient at a spin location $G_z = 430$ kT/m.

- The maximum magnetic force on CT $\langle F_x \rangle_{max} = |G_z| \mu_B = 4$ aN.

- Temperature $T = 200$ mK.

We now estimate the CT frequency shift in the spirit of the mean field approximation. Let the spin be antiparallel to the effective field $\vec{B}_{eff}$. Then

$$\frac{\langle S_z \rangle}{S} = -\frac{\langle B_{eff} \rangle_z}{B_{eff}}, \qquad (9.3)$$

where $\vec{B}_{eff} = (B_1, 0, -G_z\langle x_c\rangle)$. The net force on the CT is given by

$$F_x = -k_c\langle x_c\rangle + \gamma\hbar G_z\langle S_z\rangle. \tag{9.4}$$

Combining these formulas and averaging over fast oscillations ($\langle x_c\rangle^2 \to A^2/2$), we obtain the expressions for the relative shift of the effective spring constant and the relative frequency shift:

$$\delta k_c = -\frac{\gamma\hbar G_z^2}{[2(G_z^2 A^2 + B_1^2)]^{1/2}},$$

$$\frac{\delta\omega_c}{\omega_c} = \frac{\delta k_c}{2k_c}, \tag{9.5}$$

which corresponds to the numerical value $\delta\omega_c/\omega_c = -4.7 \times 10^{-7}$ and $\delta\omega_c = 3$ mHz. For our values of parameters, $GA \gg B_1$, and the expression for δk_c can be simplified to

$$\delta k_c = -\frac{\sqrt{2}|G_z|\mu_B}{A}. \tag{9.6}$$

Correspondingly, we have

$$\frac{\delta\omega_c}{\omega_c} = -\frac{|G_z|\mu_B}{\sqrt{2}k_c A}. \tag{9.7}$$

This new expression has a clear physical meaning. The magnetic force on the CT cannot be greater than $|G_z|\mu_B$. This is why the shift of the CT spring constant and the frequency shift decreases with the increase of the amplitude A.

Now we discuss the possibility of reducing the CT amplitude and increasing the CT frequency shift. The condition for the full adiabatic reversals can be represented as follows:

$$1 \ll \frac{|G_z|A}{B_1} \ll \frac{\omega_R}{\omega_c}. \tag{9.8}$$

The left inequality is the condition for full spin reversals (approximately between $+z$ and $-z$-directions). The right inequality is the condition for adiabatic spin motion, which is equivalent to Eq. (2.27). For our parameters

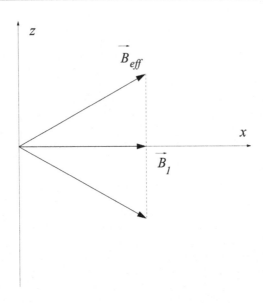

Figure 9.2: Partial reversals of the effective field in the RSC.

$|G_z|A/B_1 = 14$ and $\omega_R/\omega_c = 1270$. To increase the CT frequency shift, we may sacrifice the full spin reversals retaining the adiabatic motion. Figure 9.2 shows the partial reversals of the effective field. The use of partial adiabatic reversals is convenient for computer simulations because it allows one to save computational time. Below we show that this idea is not appropriate for the experiment, as the thermal frequency noise also increases with the decrease of the CT amplitude. While the spin is parallel or antiparallel to the effective field, the main manifestation of the CT-environment interaction in OSCAR is the thermal frequency noise. Now we will estimate its value. We will assume the "natural" detection bandwidth $\Delta f = \omega_c/4Q$ (see Chap. 7). In this case

$$x_{rms} = \left(\frac{k_B T}{k_c}\right)^{1/2}, \tag{9.9}$$

and

$$F_{rms} = \frac{k_c x_{rms}}{Q}.$$

To estimate the characteristic thermal fluctuations of the spring constant δk_c^T, we assume that the "thermal" force increases from 0 to F_{rms} when the CT coordinate changes from 0 to $A/2$. Then $\delta k_c^T = 2F_{rms}/A$, and, correspondingly, the characteristic thermal frequency fluctuations $\delta \omega_c^T$ become

$$\frac{\delta \omega_c^T}{\omega_c} = \frac{x_{rms}}{AQ}. \tag{9.10}$$

The corresponding numerical values are:

$$\begin{aligned} x_{rms} &= 68 \text{ pm}, \\ F_{rms} &= 1.6 \text{ aN}, \\ \delta \omega_c^T/\omega_c &= 1.4 \times 10^{-7}. \end{aligned} \tag{9.11}$$

Thus, the estimated characteristic CT thermal frequency fluctuations $\delta \omega_c^T$ are smaller than the absolute value of the OSCAR frequency shift $|\delta \omega_c|$. On the other hand, one can see that the thermal frequency fluctuations, like the OSCAR frequency shift, increase with the decrease of the CT amplitude. Thus, the partial adiabatic reversals will not increase the signal-to-noise ratio. Next, we consider the effect of the spin-environment interaction. This interaction can be described in terms of magnetic noise acting on the spin. Roughly speaking, this noise causes a deviation of the spin from the effective field. This deviation generates two CT trajectories corresponding to the two possible directions of the spin relative to the effective field. These two trajectories manifest the formation of the Schrödinger cat state. Now, the CT-environment interaction comes into play. CT-environment interaction quickly destroys the Schrödinger cat state leaving only one of the two possible trajectories. Physically, this appears as a quantum collapse. Usually, the collapse pushes the spin back to the "pre-collapse" direction relative to the effective field. Sometimes the spin changes its direction. When a change occurs, we can observe the quantum jump by measuring the sharp change of the CT frequency shift.

Let us assume that the collapse occurs when the separation between the two CT trajectories is of the order of the quantum uncertainty δx_c of the CT position (position of the center of mass of the ferromagnetic particle):

$$\delta x_c \sim \left(\frac{\hbar\omega_c}{k_c}\right)^{1/2}.$$ (9.12)

(In this estimate we used the first equation (4.10) omitting the factor $1/\sqrt{2}$.) If our assumption were correct then the characteristic collapse time t_{col} would be of the order of the CT period $t_{col} \sim T_c$.

Now, we will estimate the characteristic time interval between two quantum jumps t_{jump}. We assume that the most important source of the magnetic noise for the spin is associated with the cantilever modes whose frequencies are close to the Rabi frequency of the spin. The reason is the following. When the spin changes its direction between $+z$ and $-z$, its frequency in the RSC frame changes between its maximum value ω_{max} and its minimum value, which is the Rabi frequency, ω_R. Because all cantilever modes have the same thermal energy $k_B T/2$, the thermal amplitude of the mode is inversely proportional to its frequency. Thus, the greatest amplitude of the CT thermal vibrations is associated with the modes near the Rabi frequency. As an estimate, we consider those modes in the interval between the Rabi and twice the Rabi frequency ω_R and $2\omega_R$. The CT thermal amplitude A_R^T near the Rabi frequency ω_R can be estimated using the equipartition theorem:

$$\frac{1}{2}m^*\left(A_R^T\omega_R\right)^2 = k_B T.$$ (9.13)

Using the equation $\omega_c^2 = k_c/m^*$ we obtain,

$$A_R^T = \frac{\omega_c}{\omega_R}\left(\frac{2k_B T}{k_c}\right)^{1/2} \approx 75 \text{ fm}.$$ (9.14)

We will estimate the characteristic fluctuation time for the magnetic noise near the Rabi frequency as the Rabi period T_R. Then the characteristic angular deviation $\Delta\theta_0$ of the average spin $\langle\vec{S}\rangle$ during the correlation time can be estimated as

$$\Delta\theta_0 \sim \gamma(|G_z|A_R^T)T_R,$$ (9.15)

where $|G_z|A_R^T$ is the characteristic noise field near the Rabi frequency. For our parameters we obtain $\Delta\theta_0 \sim 6.8 \times 10^{-4}\ rad$. In a single reversal of the effective field the spin precession frequency in the RSC is smaller than $2\omega_R$ during the time interval Δt_1,

$$\Delta t_1 \approx 2\sqrt{3}\frac{\omega_R}{\gamma\omega_c|G_z|A}. \tag{9.16}$$

This expression can be derived from the equation:

$$\gamma B_{eff}(t) = 2\omega_R, \tag{9.17}$$

where

$$B_{eff}(t) = [(|G_z|A\cos\omega_c t)^2 + B_1^2]^{1/2}. \tag{9.18}$$

From Eq. (9.17) we obtain two solutions t_1 and t_2, which satisfy the equations

$$\cos(\omega_c t_{1,2}) = \pm\frac{\sqrt{3}\omega_R}{\gamma|G_z|A}, \tag{9.19}$$

$$\Delta t_1 = t_2 - t_1.$$

Next, subtracting the left and right sides of the equation for t_1 from the corresponding sides of the equation for t_2, and approximating $\sin(\omega_c\Delta t/2)$ with $\omega_c\Delta t/2$ we obtain (9.16).

Assuming a diffusion process, we can estimate the square of the angular deviation during a single reversal

$$\langle\Delta\theta_1^2\rangle = \frac{\Delta\theta_0^2}{T_R}\Delta t_1, \tag{9.20}$$

where $\Delta\theta_0^2/T_R$ is the diffusion coefficient. The angular deviation $\Delta\theta_{col}$ between the two collapses is

$$\langle\Delta\theta_{col}^2\rangle = \langle\Delta\theta_1^2\rangle\frac{t_{col}}{T_c/2}. \tag{9.21}$$

Now we should find the probability of a quantum jump, after the Schrödinger cat state collapse, in terms of $\langle\Delta\theta_{col}^2\rangle$. In Chap. 3 we have shown that the wave function of a spin, which points in the direction of a unit vector $\vec{n}$ can be represented by the first equation in (3.7). This equation can be simplified if we use the polar θ_n and azimuthal ϕ_n angles of the unit vector $\vec{n}$:

$$\chi_{1/2} = \cos \frac{\theta_n}{2} \, \alpha + \sin \frac{\theta_n}{2} \, \exp\left(i\phi_n\right) \beta. \tag{9.22}$$

If one measures the spin z-component, then the spin wave function $\chi_{1/2}$ collapses to the wave function α with the probability $\cos^2 \theta_n/2$, or to the wave function β with the probability $\sin^2 \theta_n/2$.

In our case we may expect a similar effect. If between two consecutive collapses the average spin deviates from the effective field by the characteristic angle $\Delta\theta_{col}$, then after the collapse it will return to its initial direction relative to the effective field with the probability $\cos^2(\Delta\theta_{col}/2)$. The probability of a quantum jump P_{jump} will be

$$P_{jump} = \sin^2 \frac{\Delta\theta_{col}}{2} \simeq \frac{\langle \Delta\theta_{coll}^2 \rangle}{4}. \tag{9.23}$$

The characteristic number of collapses between the two consecutive quantum jumps is equal to t_{jump}/t_{col}. To estimate the value of t_{jump} we may put

$$P_{jump} \frac{t_{jump}}{t_{col}} \approx 1. \tag{9.24}$$

From this equation we find

$$t_{jump} \approx \frac{t_{col}}{P_{jump}} \sim \frac{A}{\sqrt{3}\gamma|G_z|(A_R^T)^2} \approx 14 \text{ s.} \tag{9.25}$$

Note that the collapse time t_{col} cancels out in the final expression (9.25).

9.2   Experimental detection of a single spin

The first experimental detection of a single spin has been demonstrated with vitreous silica (silicon dioxide). The gamma ray irradiation produced silicon dangling bonds, which are called "E′-centers". An E′-center contains an unpaired electron spin 1/2. The estimated concentration of E'-centers was between 10^{19} and 10^{20} m^{-3}. The experiment was performed at temperature $T = 1.6$ K, with an rf field of amplitude $B_1 = 300$ μT and frequency $\omega/2\pi = 2.96$ GHz. The magnitude B_0 of the permanent magnetic field on the spin at the equilibrium CT position was

$$B_0 = |\vec{B}_{ext} + \vec{B}_d^{(0)}| = \frac{\omega}{\gamma} = 106 \text{ mT}, \tag{9.26}$$

while the external magnetic field $B_{ext} = 30$ mT. Thus, the dipole magnetic field produced by the CT was greater than the external field. The CT, with the spring constant $k_c = 110$ μN/m, oscillated parallel to the sample surface, as it is shown in Fig. 9.1, with the frequency $\omega_c/2\pi \simeq 5.5$ kHz and amplitude $A \simeq 16$ nm.

To increase the measurement sensitivity, Rugar and his team implemented a modified technique which is called the "interrupted OSCAR" technique. They interrupted the rf field periodically (with a period T_i of about 11.6 ms). When the CT was at its end point, the applied rf field was interrupted for a time interval equal to half of the CT vibration period. At the end of the "dead interval", the effective field reverses while the spin retains its initial direction. This effect is equivalent to the application of the effective π-pulse in the RSC. As a result, the CT frequency shift becomes a periodic function of time with twice the interruption period $2T_i$. Now the OSCAR signal is detected at the frequency $1/(2T_i)$. Figure 9.3 explains the interrupted OSCAR technique. One can see that the periodic interruption of the rf field with period T_i causes the periodic change of the CT frequency shift with period $2T_i$. The first harmonic of the periodic frequency shift $\delta\omega_c(t)$ is

$$\frac{4}{\pi} \delta\omega_0 \, a(t) \sin\left(\frac{\pi t}{T_i}\right). \tag{9.27}$$

Here $a(t) = \pm 1$ is a random telegraph function associated with the quantum jumps (spin flips) and $\delta\omega_0$ is the maximum value of $|\delta\omega_c|$. This harmonic has been selected for detection. In a 1 Hz detection bandwidth the frequency noise was about 23 mHz, much greater than the expected value of $\delta\omega_0$. For

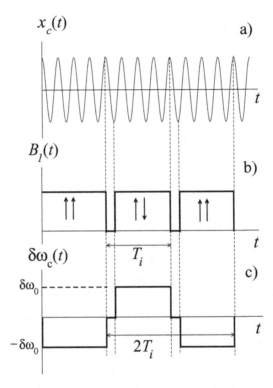

Figure 9.3: The interrupted OSCAR technique: (a) the CT coordinate as a function of time. (b) amplitude of the rf field, (c) the CT frequency shift. On panel (b) two arrows of the same direction indicate the same direction of the effective field $\vec{B}_{eff}$ and the spin magnetic moment $\vec{\mu}$, while arrows with opposite directions indicated opposite direction of $\vec{B}_{eff}$ and $\vec{\mu}$.

a spin detection, an averaging time of 13 hours per point was needed. As the averaging time was much greater than the expected characteristic time t_{jump} between the two consecutive quantum jumps, the amplitude of the first harmonic $(4/\pi)\delta\omega_0\, a(t)$ averages to zero. That is why the square of the amplitude $(4/\pi)^2\delta\omega_0^2$ has been averaged and measured experimentally instead of the amplitude itself. The maximum frequency shift was found to be $\delta\omega_0 = 4.2$ mHz.

Chapter 10

CT-Spin Dynamics in the OSCAR Technique

In this chapter we present both the quasiclassical and quantum description of the CT-spin dynamics in the OSCAR technique (Berman *et al.* [29 − 31]). In particular, we will derive the expression for the CT frequency shift $\delta\omega_c$, which is to be measured in the OSCAR experiments.

10.1 Quasiclassical theory: simple geometry

In this section we describe the quasiclassical theory of the OSCAR technique, for a simple experimental setup, shown in Fig. 10.1: the CT oscillates perpendicular to the sample surface along the z-axis, and the spin is also located on the z-axis. As we already mentioned in the previous chapter, the amplitude A of the CT vibrations in the OSCAR technique is kept constant by a feedback loop.

In order to describe the OSCAR dynamics theoretically one can use three simple approaches:

1. One can consider the CT vibrations driven by the resonant external force.

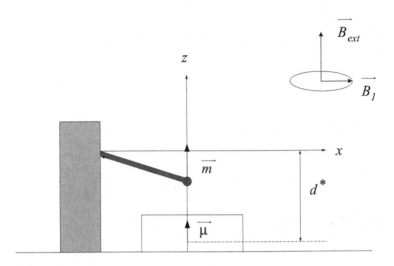

Figure 10.1: OSCAR MRFM setup. We show the case when the "up" direction of the spin magnetic moment $\vec{\mu}$ corresponds to the bottom position of the CT vibrations.

2. One can consider the free (natural) CT vibrations with no damping or during the time interval $t \ll Q/\omega_c$.

3. In numerical simulations one can consider the damped CT vibrations changing the CT coordinate at one of the end points of vibrations to the fixed value A.

In this section we will use the first approach.

The classical equation of motion for the CT (i.e. for the center of the spherical ferromagnetic particle) we will write in the form

$$\ddot{z}_c + \omega_c^2 z_c + \frac{3\mu_0 m}{2\pi m^*} \frac{\mu_z}{(d^* + z_c)^4} + \frac{\omega_c}{Q}\dot{z}_c = \frac{F_0}{m^*} \cos(\nu t + \theta_0), \qquad (10.1)$$

where the third term in the left side of the equation describes the magnetic force, produced by the spin magnetic moment $\vec{\mu}$ on the magnetic moment

$\vec{m}$ of the ferromagnetic particle, the fourth term describes the CT relaxation with time constant Q/ω_c, and the right side of the equation describes the external oscillating force of amplitude F_0 and frequency ν, which is close to the CT frequency ω_c, d^* is the distance between the spin and the origin.

As the unit of length we will use the amplitude A of the stationary CT vibrations with no spin,

$$A = \frac{F_0 Q}{m^* \omega_c^2}. \tag{10.2}$$

Also we use μ_B as the unit of the magnetic moment. Then we can write the coupled system of dimensionless equations of motion for the CT-spin system:

$$\ddot{z}_c + z_c + \frac{\lambda \mu_z}{(1 + p^* z_c)^4} + \frac{1}{Q} \dot{z}_c = \frac{1}{Q} \cos[(1 + \rho)\tau + \theta_0],$$

$$
\begin{aligned}
\dot{\mu}_x &= -\chi \, z_c \, \mu_y, \\
\dot{\mu}_y &= \epsilon \, \mu_z + \chi \, z_c \, \mu_x, \\
\dot{\mu}_z &= -\epsilon \, \mu_y.
\end{aligned}
\tag{10.3}
$$

Here we use the dimensionless time $\tau = \omega_c t$ and parameters:

$$
\begin{aligned}
p^* &= \frac{F_0 Q}{d^* m^* \omega_c^2}, \\
\lambda &= \frac{3\mu_0 m \mu_B}{2\pi (d^*)^4 Q F_0}, \\
\chi &= \frac{3\gamma \mu_0 m Q F_0}{2\pi m^* \omega_c^3 (d^*)^4}, \\
\rho &= \frac{\nu}{\omega_c} - 1, \qquad \epsilon = \frac{\omega_R}{\omega_c}.
\end{aligned}
\tag{10.4}
$$

The equations of motion of the spin magnetic moment are written in the RSC, the parameter λ describes the magnetic force on the CT, the parameter p^* describes the direct nonlinear contribution to the magnetic force, and ρ describes the deviation of the external force frequency ν from the unperturbed CT frequency ω_c. All three parameters are supposed to be small: $\lambda, p^*, \rho \ll 1$. The parameters χ and ϵ describe the z- and the x-components of the effective field on the spin. In the conditions of full adiabatic reversals we have $\chi \gg \epsilon \gg 1$.

We are going to find the CT frequency shift $\delta\omega_c$. For this we will find the value of ρ, which corresponds to the maximum CT amplitude. Let suppose that the spin magnetic moment $\vec{\mu}$ points in the positive z-direction, when the CT is at the bottom point of its stationary vibrations, (see Fig. 10.1). When the CT is at the bottom point, the dipole field B_d on the spin, is greater than its equilibrium value $B_d^{(0)}$. So, the direction of the effective field $\vec{B}_{eff}$ is close to the positive z-direction, and, consequently, the direction of $\vec{\mu}$ is close to the direction of the effective field. It is clear from Fig. 10.1 that the magnetic force on the CT (magnetic attraction, produced by the spin magnetic moment $\vec{\mu}$) is opposite to the elastic force. Thus, in this case we expect the negative frequency shift of the CT vibrations.

In order to simplify our computations we assume that the spin magnetic moment points exactly in the direction of the effective field. Assuming the "perfect" adiabatic motion for the spin we put $\dot{\vec{\mu}} = 0$ and obtain from Eqs. (10.3):

$$
\begin{aligned}
\mu_x &= \frac{\epsilon}{\sqrt{\epsilon^2 + \chi^2 z_c^2}}, \\
\mu_y &= 0, \\
\mu_z &= -\frac{\chi z_c}{\sqrt{\epsilon^2 + \chi^2 z_c^2}}.
\end{aligned}
\tag{10.5}
$$

Substituting Eqs. (10.5) into the first equation in Eqs. (10.3), we obtain the nonlinear equation for z_c:

$$
\ddot{z}_c + z_c - \frac{\lambda \chi z_c}{\sqrt{\epsilon^2 + \chi^2 z_c^2}} + \frac{1}{Q}\dot{z}_c = \frac{1}{Q}\cos[(1+\rho)\tau + \theta_0],
\tag{10.6}
$$

where we neglected the direct nonlinear contribution to the magnetic force ($p^* = 0$). Note that the third term in the left side of Eq. (10.6) corresponds to the modification of the CT potential energy by the value:

$$
-\frac{\lambda}{\chi}\sqrt{\epsilon^2 + \chi^2 z_c^2}.
\tag{10.7}
$$

The stationary solution of Eq. (10.6) can be approximately written as

$$z_c(\tau) = a(\rho) \sin[(1 + \rho)\tau + \theta_0].$$ (10.8)

Here $a(\rho)$ is the actual amplitude of of the CT vibrations, which can be slightly different from A, and we assume that z_c^2 in the denominator of the third term in the left side of Eq. (10.6) can be replaced by $1/2$:

$$z_c^2 \approx \sin^2 \tau = \frac{1}{2}(1 - \cos 2\tau) \rightarrow \frac{1}{2}.$$ (10.9)

In a rough approximation we can ignore the nonresonant term $\cos 2\tau$. In this approximation Eq. (10.6) transforms into:

$$\ddot{z}_c + \left(1 - \frac{\lambda\chi}{\sqrt{\epsilon^2 + \chi^2/2}}\right) z_c + \frac{1}{Q}\dot{z}_c = \frac{1}{Q}\cos[(1 + \rho)\tau + \theta_0].$$ (10.10)

The value of ρ corresponding to the maximum amplitude $a(\rho)$ is equal to

$$\rho = -\left(\frac{1}{4Q^2} + \frac{\lambda\chi}{2\sqrt{\epsilon^2 + \chi^2/2}}\right).$$ (10.11)

The first term in this expression describes the relative frequency shift associated with the finite quality factor of the oscillator and the second term describes the relative frequency shift $\delta\omega_c/\omega_c$ caused by the spin. Ignoring ϵ^2 compared to χ^2 we obtain a simple equation

$$\frac{\delta\omega_c}{\omega_c} = -\frac{\lambda}{\sqrt{2}}.$$ (10.12)

Taking into account that in the setup, shown in Fig. 10.1,

$$G = \frac{\partial B_d}{\partial z} = \frac{3\mu_0 m}{2\pi(d^*)^4}.$$ (10.13)

We can rewrite Eq. (10.12) in the form

$$\frac{\delta\omega_c}{\omega_c} = -\frac{\mu_B G}{\sqrt{2}k_c A}.$$ (10.14)

This equation is similar to Eq. (9.7), which was derived for the CT oscillating parallel to the sample surface.

In order to obtain a more accurate formula for the frequency shift we will use the perturbation theory developed by Bogoliubov and Mitropolsky [32]. We look for the solution of Eq. (10.6) in the form

$$z_c = a(\tau)\cos(\psi) + \lambda u_1(a, \psi), \qquad (10.15)$$

where $\psi = (1 + \rho)\tau + \theta(\tau)$. The function $u_1(a, \psi)$ is the sum of the Fourier terms with the phases 3ψ, 5ψ, 7ψ,.... The amplitudes of these terms decrease with increasing Fourier number, n, as $1/(2n + 1)^3$. The first nonvanishing term is small and equals $u_1(a, \psi) \approx 0.02\cos(3\psi)$. This allows us to neglect the contribution of $u_1(a, \psi)$ in the expression for z_c in Eq. (10.15).

The slow varying amplitude, $a(\tau)$, and the phase, $\theta(\tau)$, in the first order of the perturbation theory satisfy the two coupled differential equations,

$$\frac{da}{d\tau} = -\frac{\lambda}{2\pi}\int_0^{2\pi} d\psi \; \frac{\chi a \cos\psi \sin\psi}{\sqrt{\epsilon^2 + (\chi a \cos\psi)^2}} - \frac{a}{2Q} - \frac{\sin\theta}{Q(2 + \rho)}, \qquad (10.16)$$

and

$$\frac{d\theta}{d\tau} = -\frac{1}{8Q^2} - \rho - \frac{\lambda}{2\pi a}\int_0^{2\pi} d\psi \; \frac{\chi a \cos^2\psi}{\sqrt{\epsilon^2 + (\chi a \cos\psi)^2}} - \frac{\cos\theta}{aQ(2 + \rho)}. \qquad (10.17)$$

Note that the integral on the right-hand side of Eq. (10.16) is equal to zero. The integral on the right-hand side of Eq. (10.17) can be expressed through the elliptic integrals as

$$4\int_0^{2\pi} d\psi \; \frac{\chi a \cos^2\psi}{\sqrt{\epsilon^2 + (\chi a \cos\psi)^2}} = 4\left[\frac{1}{k}E(k) - p^2 k K(k)\right], \qquad (10.18)$$

where $k = 1/\sqrt{1 + p^2}$ and $K(k)$ and $E(k)$ are the complete elliptic integrals, respectively, of the first and second kind , $p = \epsilon/(a\psi)$. Indeed, we have

$$\int_0^{\pi/2} d\psi \; \frac{\chi a \cos^2\psi}{\sqrt{\epsilon^2 + (\chi a \cos\psi)^2}} = \int_0^{\pi/2} d\psi \; \frac{\cos^2\psi}{\sqrt{p^2 + \cos^2\psi}} \qquad (10.19)$$

$$= \int_0^{2\pi} d\psi \; \frac{(1 - \sin^2\psi)}{\sqrt{p^2 + 1 - \sin^2\psi}} = \int_0^{\pi/2} d\psi \; \frac{(p^2 + 1)\left(1 - \frac{1}{p^2+1}\sin^2\psi\right) - p^2}{\sqrt{p^2 + 1}\sqrt{1 - \frac{1}{p^2+1}\sin^2\psi}}.$$

Splitting this integral in two parts, we obtain the right-hand side of Eq. (10.18).

When $p^2 \ll 1$, one can decompose $K(k)$ and $E(k)$ as

$$K(k) \approx C + (C-1)\frac{k'^2}{4} + \ldots,$$

$$E(k) \approx 1 + \left(C - \frac{1}{2}\right)\frac{k'^2}{2} + \ldots,$$

(10.20)

where $k'^2 = 1 - k^2 \approx p$, $C = \ln(4/k') \approx \ln(4/p)$. From Eqs. (10.18) and (10.20), we find the value of the integral in Eq. (10.17) for $p \ll 1$

$$-\frac{\lambda}{2\pi a}\int_0^{2\pi} \frac{\chi a \cos^2 d\psi}{\sqrt{\epsilon^2 + (\chi a \cos\psi)^2}} \approx -\frac{2\lambda}{\pi a}\left[1 - \frac{p^2}{a}\left(2\ln\frac{4}{p} - 1\right)\right].$$

(10.21)

Substituting Eq. (10.21) into Eq. (10.17), we obtain

$$\frac{da}{d\tau} = -\frac{a}{2Q} - \frac{1}{Q(2+\rho)}\sin\theta,$$

$$\frac{d\theta}{d\tau} = -\frac{1}{8Q^2} - \rho - \frac{2\lambda}{\pi a}\left[1 - \frac{p^2}{4}\left(2\ln\frac{4}{p} - 1\right)\right] - \frac{\cos\theta}{aQ(2+\rho)}.$$

(10.22)

We now calculate the position of the maximum of the amplitude, $a(\rho)$, in the stationary regime of driven oscillations using Eq. (10.22). In the regime of driven oscillations $a = $ const, $\theta = $ const, and we must solve the system of two Eqs. (10.22) where $da/d\tau = 0$ and $d\theta/d\tau = 0$. Eliminating the phase θ, we have

$$\frac{1}{a^2(2+\rho)^2} = \frac{1}{4} + Q^2\left(\frac{1}{8Q^2} + \rho + \frac{2\lambda}{\pi a}\right)^2,$$

(10.23)

where we neglected the term proportional to $p^2 \ll 1$. The amplitude, a, can be written as $a = 1 + b$, where $b \ll 1$, so that

$$\frac{1}{a(2+\rho)} = \frac{1}{(1+b)(2+\rho)} \approx \frac{1}{2}\left(1 - b - \frac{\rho}{2}\right).$$

(10.24)

Taking the square root from both side of Eq. (10.23) and using Eq. (10.24), we obtain

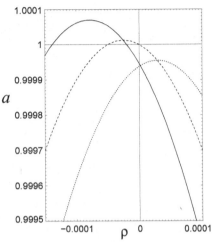

Figure 10.2: The dependence of the amplitude a of the driven CT oscillations on the frequency detuning, ρ, obtained using the numerical solution of the equations of motion (10.3). The solid line corresponds to the initial conditions (10.28) and the values of parameters $\lambda = 8.5 \times 10^{-5}$, $\chi = 2500$, $\epsilon = 280$ $p = 0.05$, $Q = 100$. The dotted line corresponds to the same values of the parameters but for "inverted" initial condition (in Eq. (10.28) $\mu_x \to -\mu_x$, $\mu_z \to -\mu_z$). The dashed line represents the dependence $a(\rho)$ with no spin.

$$-b - \frac{\rho}{2} \approx 2Q^2 \left(\frac{1}{8Q^2} + \rho + \frac{2\lambda}{\pi} \right)^2, \qquad (10.25)$$

where we put $a \approx 1$ in the denominator of the term proportional to λ (i.e. we neglected the term of the order of $\beta\lambda$.

The maximum of the function, $\beta = \beta(\rho)$ can be found from the condition $d\beta/d\rho = 0$, which yields

$$\rho = -\frac{1}{4Q^2} - \frac{2\lambda}{\pi}. \qquad (10.26)$$

Thus, the relative frequency shift caused by the spin is given by

$$\frac{\delta\omega_c}{\omega_c} = -\frac{2\lambda}{\pi}. \qquad (10.27)$$

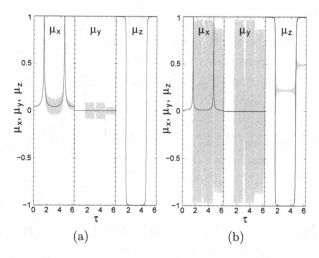

(a)   (b)

Figure 10.3: (a) Dynamics of the magnetic moment $\vec{\mu}(\tau)$, with the initial conditions (10.28). The gray line is obtained as a result of the numerical integration of Eqs. (10.3), while the black one indicates the solution (10.5). For $\mu_z(\tau)$ both curves almost coincide. The parameters are the same as those for the solid line in Fig. 10.2. (b) The same as (a) but for $\epsilon = 28$.

Comparing with the formula (10.12) we see that the more accurate perturbation approach yields the factor $2/\pi$ instead of the factor $1/\sqrt{2}$.

The numerical solution of the original equations (10.3) confirms the analytical estimates. Figure 10.2 (solid line) demonstrates the dependence of the stationary amplitude of the CT vibrations, a, on the frequency detuning, ρ. The stationary amplitude is achieved at $\tau \gg Q$. The initial conditions in this numerical solution correspond to the magnetic moment pointing in the direction of the effective field:

$$
\begin{aligned}
z_c(0) &= -1, \quad \dot{z}_c(0) = 0, \\
\mu_x(0) &= \frac{\epsilon}{\sqrt{\epsilon^2 + \chi^2}}, \quad \mu_y(0) = 0, \\
\mu_z(0) &= \frac{\chi}{\sqrt{\epsilon^2 + \chi^2}}, \quad \theta_0 = \frac{3}{2}\pi.
\end{aligned}
\tag{10.28}
$$

The dotted line on the same Figure corresponds to the magnetic moment pointing opposite to the direction of the effective field. One can see that the CT frequency caused by the spin, changes its sign.

Figure 10.3(a) demonstrates an adiabatic reversal of the magnetic moment $\vec{\mu}$ obtained from the adiabatic approximation (10.5) (black line) and original equations of motion (10.3) (gray line). One can see a close correspondence between the two solutions.

For comparison, Fig. 10.3(b) demonstrates the dependence $\vec{\mu}(\tau)$ for the case when the conditions of adiabatic reversals are violated. In this case the adiabatic approximation is not close to the solution of Eq. (10.3).

10.2 Quantum theory of the OSCAR MRFM

In this section we use the quantum theory to describe the free vibrations of the CT interacting with a single spin. We will consider the same simple setup in Fig. 10.1 as in the previous section. We will use the quantum units for the energy, coordinate and momentum, and the quantum Hamiltonian (8.4) of the CT-spin system, where we put $\dot{\phi} = 0$. We will describe the computer simulations of the quantum dynamics for the values of parameters

$$\eta = 0.3, \quad \epsilon = 10, \quad \langle z_c(0) \rangle = A = 13, \quad \langle p_c(0) \rangle = 0. \qquad (10.29)$$

The condition for the adiabatic motion in terms of our parameters is

$$2\eta A \ll \epsilon^2. \qquad (10.30)$$

The condition for the full reversals of the effective field is

$$\epsilon \ll 2\eta A. \qquad (10.31)$$

For the parameters (10.29), we have $2\eta A \approx 7.8$. Thus, the condition (10.30) is satisfied while (10.31) is violated. As we have explained in Sec. 1 of Chap. 9 for such values of parameters we have the case of partial adiabatic reversals shown in Fig. 9.2, which is convenient for computer simulations. In order to estimate the CT frequency shift, caused by the spin, in the conditions of the partial adiabatic reversals we may use formula (10.11). In units of ω_c the value of $\delta\omega_0 = |\delta\omega_c|$ is

$$\delta\omega_0 = \frac{\lambda\chi}{2\sqrt{\epsilon^2 + \chi^2/2}}. \tag{10.32}$$

Taking into consideration that the parameters λ and χ can be rewritten in the form:

$$\chi = \frac{\gamma A}{\omega_c}G, \quad \lambda = \frac{\mu_B}{k_c A}G, \tag{10.33}$$

and also $\mu_B = \gamma\hbar/2$, and the quantum unit of length is $(\hbar\omega_c/k_c)^{1/2}$, we obtain

$$\lambda\chi = 2\eta^2, \quad \chi^2 = 4\eta^2 A^2. \tag{10.34}$$

Substituting Eqs. (10.34) into Eq. (10.32), for the values of parameters (10.29), we obtain

$$\delta\omega_0 = \frac{\eta^2}{\sqrt{2\eta^2 A^2 + \epsilon^2}} \approx 7.9 \times 10^{-3}. \tag{10.35}$$

The results of computer simulations of the OSCAR MRFM dynamics are similar to those described in Chap. 8, but with some important peculiarities. The wave functions was taken in the form (8.7) with the initial conditions (8.11). The complex parameter α in Eq. (8.11) is equal to

$$\alpha = \frac{1}{\sqrt{2}}\left[\langle z_c(0)\rangle + i\langle p_c(0)\rangle\right], \tag{10.36}$$

where $\langle z_c(0)\rangle$ and $\langle p_c(0)\rangle$ are taken from (10.29). The probability distribution for the CT position

$$P(z_c, \tau) = |\psi_1(z_c, \tau)|^2 + |\psi_2(z_c, \tau)|^2, \tag{10.37}$$

eventually splits into two peaks which describe two CT trajectories.

The ratio of the integrated probabilities for two peaks is given approximately by $\tan^2(\theta/2)$, where θ is the initial angle between the directions of the average spin and the effective field $\vec{B}_{eff} = (\epsilon, 0, -2\eta\langle z_c(0)\rangle)$. (The effective

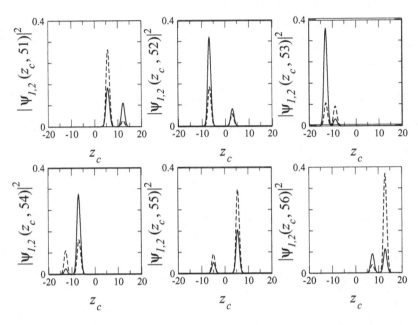

Figure 10.4: Functions $|\psi_1(z_c, \tau)|^2$ (solid curve) and $|\psi_2(z_c, \tau)|^2$ (dashed curve), for six values of τ. Initial conditions: $\langle z_c(0) \rangle = A = 13$, $\langle p_c(0) \rangle = 0$, and the electron spin points in the positive z-direction.

field is given in units ω_c/γ.) The approximate position of the center of the first peak is

$$z_1 = A \cos(1 - \delta\omega_0)\tau, \tag{10.38}$$

and the approximate position of the second peak is

$$z_2 = A \cos(1 + \delta\omega_0)\tau, \tag{10.39}$$

where $\delta\omega_0 \approx 8.0 \times 10^{-3}$. Note that the value $\delta\omega_0$ estimated from (10.35) is almost the same. Both functions $\psi_1(z_c, \tau)$ and $\psi_2(z_c, \tau)$ contribute to each peak (see Fig. 10.4). When two peaks are clearly separated, the wave function Ψ can be approximately represented as a sum of two functions Ψ^b and Ψ^{sm}, which correspond to the "big" and "small" peaks in the probability distribution. We have found that each function Ψ^b and Ψ^{sm}, with the accuracy to 1%, can be represented as a product of the coordinate and spin functions

$$\Psi^b = \psi^b(z_c, \tau)\chi^b(s, \tau), \quad \Psi^{sm} = \psi^{sm}(z_c, \tau)\chi^{sm}(s, \tau). \tag{10.40}$$

The first spin function $\chi^b(s, \tau)$ describes the average spin which points approximately opposite to the direction of the effective field

$$\vec{B}_{eff,1} = (\epsilon, 0, -2\eta z_1),$$

for the first CT trajectory. The second spin function $\chi^{sm}(s, \tau)$ describes the average spin which points approximately in the direction of the effective field $\vec{B}_{eff,2} = (\epsilon, 0, -2\eta z_2)$, for the second CT trajectory.

Unlike MRFM dynamics considered in Chap. 8, the OSCAR technique implies different effective fields for two CT trajectories. That is why, in general, the average spins corresponding to two CT trajectories do not point in the opposite directions, and the wave functions $\chi^b(s, \tau)$ and $\chi^{sm}(s, \tau)$ are not orthogonal to each other. The only exceptions are the instants τ for which two CT trajectories intersect providing a unique direction for the effective field.

We would like to note the important difference between the quasiclassical and quantum dynamics of the CT-spin system. If initially the spin magnetic moment makes an angle θ with the effective field then the quasiclassical dynamics describes a single CT trajectory with the frequency shift

$$\delta\omega_c = -\delta\omega_0 \cos\theta, \tag{10.41}$$

where $\delta\omega_0$ is the maximum absolute value of the frequency shift, which can be achieved at $\theta = 0$ and $\theta = \pi$. Thus, the frequency shift $\delta\omega_c$ can accept any value between $-\delta\omega_0$ and $\delta\omega_0$ depending on the value of the angle θ. The quantum dynamics describes two CT trajectories at the same time (the Schrödinger cat state) with the frequency shifts $-\delta\omega_0$ and $\delta\omega_0$, corresponding to the two possible values of the spin component relative to the direction of the effective field. The probabilities for these two trajectories are $\cos^2\theta/2$

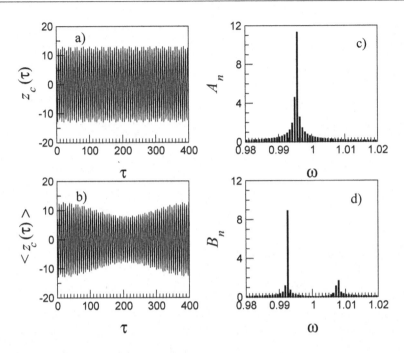

Figure 10.5: Quasiclassical and quantum CT dynamics in OSCAR MRFM; (a) classical cantilever coordinate $z_c(\tau)$, (b) quantum average cantilever coordinate $\langle z_c(\tau) \rangle$, (c) the Fourier spectrum for (a): $z_c(\tau) = \sum_n A_n \cos(\omega_n \tau + \varphi_n)$, (d) the Fourier spectrum for (b): $\langle z_c(\tau) \rangle = \sum_n B_n \cos(\omega_n \tau + \theta_n)$. All parameters are the same as in Fig. 10.4. In (a) and (b) we show, for convenience, time sequences shorter than those used in order to get the Fourier spectrum shown in (c) and (d).

and $\sin^2 \theta/2$. For any angle θ the quantum dynamics describes only two frequencies $\delta\omega_0$ and $-\delta\omega_0$. Figure 10.5 demonstrates the difference between the quasiclassical and the quantum dynamics. In order to describe the decoherence and the thermal diffusion for the CT interacting with the spin we have used the master equation (8.18), where we put $\dot{\phi} = 0$. The initial density matrix was taken in the form (8.21), which is the product of the CT and the spin density matrices with the CT being in the coherent state, and the spin pointing in the positive z-direction.

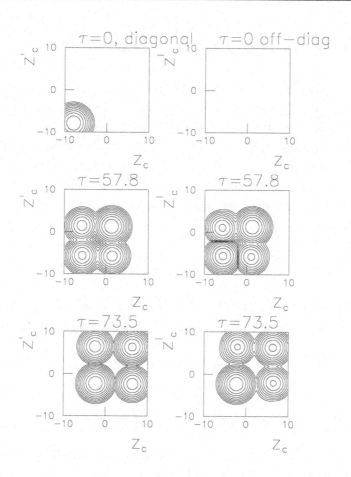

Figure 10.6: Evolution of the density matrix for ($\beta = D = 0$), at four instants of time (as indicated in the figure). Parameters: $\epsilon = 10$, $\eta = 0.3$. The left column: $\ln |\sum_s \rho_{s,s}|$, the right column: $\ln |\sum_s \rho_{s,-s}|$. Initial conditions: $\langle z_c(0) \rangle = -8$, $\langle p_c(0) \rangle = 0$. Contour lines are the intersections of the functions $\ln |\sum_s \rho_{s,\pm s}|$ horizontal planes.

Figure 10.6 shows the evolution of the density matrix without decoherence and thermal diffusion, for the following values of parameters

$$\epsilon = 10, \quad \eta = 0.3, \quad \beta = D = 0. \qquad (10.42)$$

The left column demonstrates the behavior of the spin diagonal density matrix elements (contour lines for $\ln |\sum_s \rho_{s,s}|$), and the right column demonstrates the behavior of the spin non-diagonal density matrix elements (contour lines for $\ln |\sum_s \rho_{s,-s}|$). Initially we have one "spatial" peak on the plane $z_c - z_c'$. Eventually, this peak splits into four peaks. Two spatial diagonal peaks, which are centered on the line $z_c = z_c'$, correspond to two cantilever trajectories. Two spatial non-diagonal peaks describe a "coherence" between the two trajectories, which is a quantitative characteristic of the Schrödinger cat state. All four spin components of the density matrix $\rho_{s,s'}(z_c, z_c', \tau)$ (with $s = \pm 1/2, s' = \pm 1/2$) contribute to each peak in the $z_c - z_c'$ plane.

The density matrix $\rho_{s,s'}(z_c, z_c', \tau)$ can be represented as a sum of four terms

$$\rho_{s,s'}(z_c, z_c', \tau) = \sum_{k=1}^{4} \rho_{s,s'}^{(k)}(z_c, z_c', \tau), \qquad (10.43)$$

where each term describes one peak in the $z_c - z_c'$ plane: the first two terms with $k = 1, 2$ describe the spatial diagonal peaks, and two other terms with $k = 3, 4$ describe the spatial non-diagonal peaks.

It was found that the diagonal terms $\rho^{(1)}$ and $\rho^{(2)}$ can be approximately decomposed into the product of the coordinate and spin parts

$$\rho_{s,s'}^{(k)}(z_c, z_c', \tau) = \hat{R}^{(k)}(z_c, z_c', \tau)\hat{\chi}_{s,s'}^{(k)}(\tau). \qquad k = 1, 2. \qquad (10.44)$$

The spin matrix $\hat{\chi}_{s,s'}^{(1)}(\tau)$ describes the average spin which points approximately opposite to the direction of the effective field $\vec{B}_{eff,1}$. The spin matrix $\hat{\chi}_{s,s'}^{(2)}(\tau)$ describes the average spin which points approximately in the direction of the effective field $\vec{B}_{eff,2}$.

The same properties of the density matrix remain for the case $\beta, D \neq 0$. Figure 10.7 demonstrates the effects of decoherence and thermal diffusion for $\beta = 0.001$ and $D = 20$. The spatial non-diagonal peaks in the $z_c - z_c'$ plane quickly decay. This reflects the effect of decoherence: the statistical mixture of two possible CT trajectories replaces the Schrödinger cat state. Next, the spatial diagonal peaks spread out along the diagonal $z_c = z_c'$. This reflects the classical effect of the thermal diffusion.

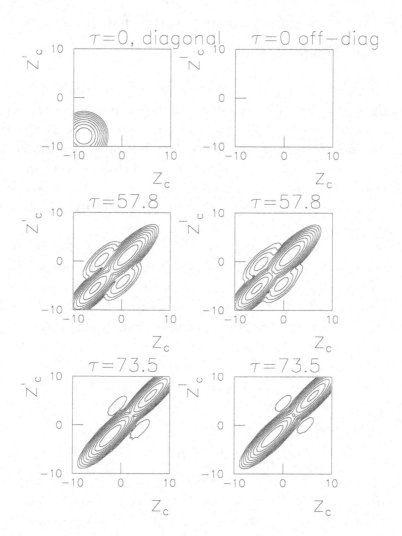

Figure 10.7: Evolution of the density matrix: effects of decoherence and thermal diffusion; $D = 20$, $\beta = 0.001$, $\epsilon = 10$, $\eta = 0.3$. The left column: $\ln |\sum_s \rho_{s,s}|$, the right column: $\ln |\sum_s \rho_{s,-s}|$. Initial conditions: $\langle z_c(0) \rangle = -8$, $\langle p_c(0) \rangle = 0$. Contour lines are the intersections of the functions $\ln |\sum_s \rho_{s,\pm s}|$ with the horizontal planes.

10.3   OSCAR frequency shift for a realistic setup

As we have mentioned in the previous chapter the first experimental detection of a single spin has been demonstrated with the setup shown in Fig. 9.1. In that experiment the dipole field $\vec{B}_d$ produced by the CT was greater than the external field $\vec{B}_{ext}$. It means that for the CT position $x_c = 0$ with no rf field the net magnetic field on the spin $\vec{B}_0 = \vec{B}_d^{(0)} + \vec{B}_{ext}$ does not point, even approximately, in the positive z direction. Moreover, the spin does not have to be located in the $x - z$ plane, so that the dipole field $\vec{B}_d^{(0)}$ may have the non-zero $y-$component.

In this section we will compute the OSCAR frequency shift for the CT oscillating parallel to the sample surface (Fig. 9.1), for arbitrary location of the spin and arbitrary relation between the dipole field $\vec{B}_d$ and the permanent external field $\vec{B}_{ext}$. First, we will assume that the rf field of frequency ω is linearly polarized in the plane which is perpendicular to the magnetic field $\vec{B}_0$. Later, we will discuss the arbitrary direction of the rf field. As only one rotating component of the rf field will be important, we will write the rf field with a factor 2: $2\vec{B}_1(t)$. We will consider the free vibrations of the CT interacting with the spin magnetic moment using the quasiclassical approach.

The dipole magnetic field $\vec{B}_d$ is given by:

$$\vec{B}_d = \frac{\mu_0}{4\pi} \frac{3(\vec{m} \cdot \vec{n})\vec{n} - \vec{m}}{\tilde{r}^3}, \tag{10.45}$$

where $\tilde{r}$ is the variable distance between the moving CT and the stationary spin, $\vec{n}$ is the unit vector pointing from the CT to the spin. We put:

$$\tilde{r} = \sqrt{(x - x_c)^2 + y^2 + z^2}, \tag{10.46}$$

$$\vec{n} = \left(\frac{x - x_c}{\tilde{r}}, \frac{y}{\tilde{r}}, \frac{z}{\tilde{r}} \right), \tag{10.47}$$

where x, y, z are the spin coordinates, and x_c is the CT-coordinate. For $x_c = 0$ and $B_1 = 0$ the net magnetic field on the spin is given by:

$$\vec{B}_0 = \vec{B}_{ext} + \vec{B}_d^{(0)}, \tag{10.48}$$

$$\vec{B}_d^{(0)} = \frac{3m\mu_0}{4\pi r^5}\left(zx, \ zy, \ z^2 - \frac{r^2}{3}\right), \tag{10.49}$$

$$\vec{B}_{ext} = (0, \ 0, \ B_{ext}), \tag{10.50}$$

where $r = \sqrt{x^2 + y^2 + z^2}$. In the linear approximation on x_c the magnetic field $\vec{B}_d$ changes by the value of $\vec{B}_d^{(1)}$:

$$\vec{B}_d^{(1)} = -(G_x, G_y, G_z)\, x_c, \tag{10.51}$$

where

$$(G_x, G_y, G_z) = \frac{3m\mu_0}{4\pi r^7}\left(zr^2 - 5zx^2, -5xyz, xr^2 - 5xz^2\right). \tag{10.52}$$

The quantities (G_x, G_y, G_z) describe the gradient of the magnetic field at the spin location at $x_c = 0$:

$$(G_x, G_y, G_z) = \left(\frac{\partial B_x}{\partial x}, \frac{\partial B_y}{\partial x}, \frac{\partial B_z}{\partial x}\right). \tag{10.53}$$

Note, that the magnetic field and its gradient depend on the CT coordinate x_c.

Next we consider the equation of motion for the spin magnetic moment $\vec{\mu}$ in the RSC rotating with the rf field of frequency ω about the magnetic field $\vec{B}_0$ (the $\tilde{z}$-axis of this new system points in the direction of $\vec{B}_0$). We have:

$$\dot{\vec{\mu}} \quad = -\gamma\vec{\mu} \times \vec{B}_{eff},$$

$$\vec{B}_{eff} \quad = \left(B_1, \ 0, \ B_0 - \frac{\omega}{\gamma} - x_c \sum_i G_i \cos\alpha_i\right), \tag{10.54}$$

$$\cos\alpha_i \quad = \frac{B_{0,i}}{B_0}, \quad i = x, y, z.$$

Here α_i, $(i = x, y, z)$ are the angles between the direction of the magnetic field $\vec{B}_0$ and the axes x, y, z of the laboratory system of coordinates. Note, that we ignore the transversal components of the dipole field $\vec{B}_d$ as they represent the fast oscillating terms in the RSC. Also we consider only rotating component of the rf magnetic field.

The classical equation of motion for the CT can be represented as:

$$\ddot{x}_c + \omega_c^2 x_c = \frac{F_x}{m^*}, \qquad (10.55)$$

where F_x is the magnetic force acting on the on the CT. Note, that we consider CT oscillations in the laboratory system of coordinates. Ignoring fast oscillating terms in the laboratory system, we obtain:

$$F_x = -\mu_{\bar{z}} \sum_i G_i \cos \alpha_i. \qquad (10.56)$$

Below, in this section, we will use the following units: for frequency — "ω_c", for magnetic moment — "μ_B", for magnetic field — "ω_c/γ", for length — the characteristic distance "L_0" between the CT and the spin, for force — "$k_c L_0$". Using these units and the dimensionless time $\tau = \omega_c t$, we derive the dimensionless equations of motion:

$$\dot{\vec{\mu}} = -\vec{\mu} \times \vec{B}_{eff},$$

$$\ddot{x}_c + x_c = F_x,$$

$$\vec{B}_{eff} = (B_1, 0, \Delta - \alpha' G' x_c),$$

$$\qquad\qquad\qquad (10.57)$$

$$F_x = -\alpha \alpha' G' \mu_{\bar{z}},$$

$$\Delta = B_0 - \omega,$$

$$G' = \frac{1}{r^7}[z(r^2 - 5x^2)\cos\alpha_x - 5xyz\cos\alpha_y + x(r^2 - 5z^2)\cos\alpha_z].$$

Parameters α and α' are given by:

$$\alpha = \frac{\mu_B \omega_c}{\gamma k_c L_0^2}, \qquad \alpha' = \frac{3\gamma \mu_0 m}{4\pi \omega_c L_0^3}. \qquad (10.58)$$

Note that all quantities in Eq. (10.57) are dimensionless, i.e. x means x/L_0, μ means μ/μ_B, B_0 means $\gamma B_0/\omega_c$, and so on. In terms of dimensional quantities the parameter α' equals to the ratio of the dipole frequency $\gamma B_d^{(0)}$ to the CT frequency ω_c, and the product $\alpha\alpha'$ equals to the ratio of the static CT displacement F_x/k_c to the CT-spin distance L_0. The derived equations are convenient for both numerical simulations and analytical estimates.

When the CT oscillates, the resonant condition $\omega = |\vec{B}_{ext} + \vec{B}_d|$ can be satisfied only if the spin is located inside the resonant slice which is defined by its boundaries:

$$|\vec{B}_{ext} + \vec{B}_d(x_c = \pm A)| = \omega, \qquad (10.59)$$

where A is the amplitude of the CT vibrations. For analytical estimate we assume that the spin is located at the central surface of the resonant slice. In this case in Eq. (10.57), $\Delta = 0$. We will assume an ideal adiabatic motion and put $\ddot{\vec{\mu}} = 0$ in Eq. (10.57). Let the CT starts its motion at $t = 0$ from the right end position $x_c(0) = A$. Then, the initial direction of the effective magnetic field $\vec{B}_{eff}(0)$ depends on the sign of the parameter G'. If $G' < 0$, then the direction of $\vec{B}_{eff}(0)$ is close to the direction of $\vec{B}_0$. If $G' > 0$, then the vector $\vec{B}_{eff}$ points, approximately, in the direction opposite to the vector $\vec{B}_0$. We will assume that the spin magnetic moment is initially in the state of equilibrium. It means that for $G' < 0$ the magnetic moment $\vec{\mu}$ will have the direction close to the direction of the effective field $\vec{B}_{eff}$.

From the first equation in (10.57), we obtain

$$\mu_z \simeq \frac{B_{eff,z} G'}{B_{eff} |G'|}. \qquad (10.60)$$

Substituting this expression into F_x we obtain the following equation for x_c:

$$\ddot{x}_c + x_c \left[1 + \frac{\alpha(\alpha')^2 G' |G'|}{\sqrt{B_1^2 + (\alpha' G' x_c)^2}} \right] = 0. \qquad (10.61)$$

We solve this equation using the perturbation theory [32]. Eq. (10.61) can be written in the following form:

$$\frac{d^2 x_c}{d\tau^2} + x_c = \epsilon' f(x_c), \tag{10.62}$$

where

$$f(x_c) = \frac{\alpha' G' x_c}{\sqrt{B_1^2 + (\alpha' G')^2 x_c^2}}, \tag{10.63}$$

and $\epsilon' = -\alpha \alpha' |G'|$.

The approximate solution of (10.62) can be written as:

$$x_c(\tau) = a(\tau) \cos \psi(\tau) + O(\epsilon'), \tag{10.64}$$

where in the first order on ϵ', the functions $a(\tau)$ and $\psi(\tau)$ satisfy the equations:

$$\frac{da}{d\tau} = \epsilon' P_1(a) + O(\epsilon'),$$

$$\frac{d\psi}{d\tau} = 1 + \epsilon' Q_1(a) + O(\epsilon'), \tag{10.65}$$

and the functions $P_1(a)$ and $Q_1(a)$ are given by:

$$P_1(a) = -\frac{1}{2\pi} \int_0^{2\pi} f(a \cos \psi) \sin \psi \, d\psi, \tag{10.66}$$

$$Q_1(a) = -\frac{1}{2\pi a} \int_0^{2\pi} f(a \cos \psi) \cos \psi \, d\psi. \tag{10.67}$$

On inserting the explicit expression (10.63) for $f(a \cos \psi)$ one gets:

$$P_1(a) = 0, \tag{10.68}$$

$$Q_1(a) = -\frac{2\alpha' G'}{\pi \sqrt{B_1^2 + (\alpha' G' a)^2}} \int_0^{\pi/2} \frac{(1 - \sin^2 \psi)}{\sqrt{1 - k^2 \sin^2 \psi}} \, d\psi, \tag{10.69}$$

where

$$k^2 = \frac{(\alpha' G' a)^2}{B_1^2 + (\alpha' G' a)^2}. \tag{10.70}$$

Equation (10.69) can be written as:

$$Q_1(a) = -\frac{2\alpha'G'}{\pi k^2 \sqrt{B_1^2 + (\alpha'G'a)^2}}[(k^2 - 1)K(k) + E(k)], \qquad (10.71)$$

where $K(k)$ and $E(k)$ are the complete elliptic integrals of the first and second kind. When $k \simeq 1$ elliptic integrals can be approximated by:

$$K(k) \approx \ln\frac{4}{\sqrt{1-k^2}} + \frac{1}{4}\left(\ln\frac{4}{\sqrt{1-k^2}} - \frac{1}{2}\right)(1-k^2),$$

$$E(k) \approx 1 + \frac{1}{2}\left(\ln\frac{4}{\sqrt{1-k^2}} - \frac{1}{2}\right)(1-k^2). \qquad (10.72)$$

In the first approximation we have $a = A$ and

$$\delta\omega_c = \frac{2}{\pi}\frac{\alpha(\alpha')^2 G'|G'|}{\sqrt{B_1^2 + (\alpha'G'A)^2}} \times$$

$$\left\{1 + \frac{1}{2}\frac{B_1^2}{B_1^2 + (\alpha'G'A)^2}\left[\ln\left(\frac{4\sqrt{B_1^2 + (\alpha'G'A)^2}}{B_1}\right) + \frac{1}{2}\right]\right\}. \qquad (10.73)$$

In typical experimental conditions we have

$$B_1 \ll \alpha'|G'|A, \qquad (10.74)$$

and Eq. (10.73) transforms to a simple expression

$$\delta\omega_c = \frac{2}{\pi}\frac{\alpha\alpha'G'}{A}. \qquad (10.75)$$

In terms of dimensional quantities we have:

$$\frac{\delta\omega_c}{\omega_c} = \frac{2\mu_B G_0}{\pi k_c A}, \qquad (10.76)$$

where

$$G_0 = \sum_i G_i \cos\alpha_i. \qquad (10.77)$$

Taking into consideration the expression for λ in (10.33) one can see that formula (10.76) for the relative frequency shift is similar to formula (10.27) which was derived for the simplest OSCAR setup. Note, that Eqs. (10.73) and (10.75)–(10.77) are valid for any spin location on the central resonant surface and for any relation between B_{ext} and B_d. We also note that in Eq. (10.73), the frequency shift $\delta\omega_c$ is an even function of y and an odd function of x.

Below we will describe the results of numerical solution of Eq. (10.57) with the parameters taken from experiment [8]:

$$\frac{\omega_c}{2\pi} = 5.5 \text{ kHz}, \quad k_c = 110 \text{ } \mu\text{N/m}, \quad A = 16 \text{ nm},$$

$$B_{ext} = 30 \text{ mT}, \quad \frac{\omega}{2\pi} = 2.96 \text{ GHz}, \quad \frac{\omega}{\gamma} = 106 \text{ mT}, \quad (10.78)$$

$$|G_z| = 200 \text{ kT/m}, \quad B_1 = 300 \text{ } \mu\text{T}.$$

Taking the characteristic distance between the spin and the CT $L_0 = 350$ nm we obtain the following values of the dimensionless parameters:

$$\alpha = 1.35 \times 10^{-13} \quad \alpha' = 1.07 \times 10^6, \quad A = 4.6 \times 10^{-2},$$

$$(10.79)$$

$$B_1 = 1.5 \times 10^3, \quad B_{ext} = 1.53 \times 10^5, \quad \omega = 5.4 \times 10^5.$$

The initial conditions were taken in the form

$$\vec{\mu}(0) = (0, 0, 1),$$

$$(10.80)$$

$$x_c(0) = A, \quad \dot{x}_c(0) = 0.$$

Figure 10.8 shows the frequency shift $\delta\omega_c$ as a function of the spin z-coordinate at the central resonant surface ($\Delta = 0$). First, one can see an excellent agreement between the numerical data and the analytical estimate (10.73). Second, as it may be expected, the maximum magnitude of the frequency shift $|\delta\omega_c|$ can be achieved when the spin is located in the plane of the

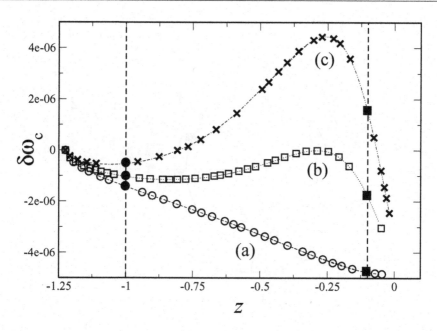

Figure 10.8: The OSCAR MRFM frequency shift $\delta\omega_c(z)$ at the central resonant surface ($\Delta = 0$), for $x > 0$. Symbols show numerical data, lines correspond to estimate (10.73) for (a) $y = 0$ (circles), (b) $y = x/2$ (squares) and (c) $y = x$ (crosses). Black circles correspond to the plane $z = -1$, black squares correspond to the plane $z = -0.1$.

CT vibrations $y = 0$. However, for $y = x$, one can achieve almost the same magnitude $|\delta\omega_c|$ (with the opposite sign of $\delta\omega_c$). Moreover, for $y = x$, the dependence $\delta\omega_c(z)$ has an extremum which can be used for the measurement of the spin z-coordinate. If the distance between the CT and the surface of the sample can be controlled, then the "depth" of the spin location below the sample surface can be determined. Note, that in Figs. 10.8–10.12, the coordinates x, y, and z are given in units of L_0, and the frequency shift is in units of ω_c.

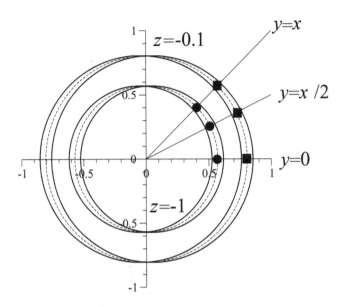

Figure 10.9: Cross-sections of the resonant slice for $z = -0.1$ and $z = -1$. Dashed lines show the intersection between the cross-sections and the central resonant surface. The black circles and squares indicate spin locations which correspond to the frequency shifts indicated by the same symbols in Fig. 10.8.

Figure 10.9 shows the cross-sections of the resonant slice for $z = -0.1$ and $z = -1$. The greater distance from the CT the smaller cross-sectional area. The black circles and squares in Fig. 10.9 show the spin locations, which correspond to the frequency shifts indicated by the same symbols in Fig. 10.8.

Figure 10.10 demonstrates the "radial" dependence of the frequency shift $\delta\omega_c(r_p)$, where $r_p = (x^2 + y^2)^{1/2}$. The value of r_p can be changed by the lateral displacement of the cantilever. As one may expect, the maximum value of $|\delta\omega_c|$ corresponds to the central resonant surface. The maximum becomes sharper with the decrease of z. Thus, the small distance between the CT and the sample surface is preferable for the measurement of the "radial position" of the spin.

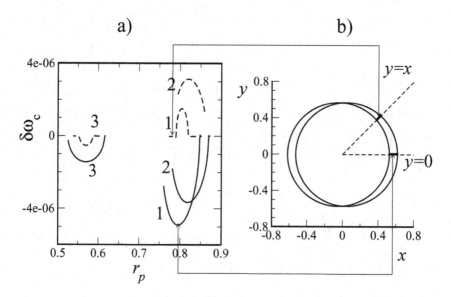

Figure 10.10: (a) The OSCAR MRFM frequency shift $\delta\omega_c(r_p)$ inside the cross-sectional area of the resonant slice for $x > 0$. Solid lines correspond to $y = 0$, dashed lines correspond to $y = x$. Lines are (1), $z = -0.1$, (2), $z = -0.43$, and (3), $z = -1$. $r_p = (x^2 + y^2)^{1/2}$. (b) Cross-section of the resonant slice for $z = -0.1$. Bold segments show the spin locations which correspond to the lines (1) on (a).

Figure 10.11 shows the "azimuthal dependence" of the frequency shift $\delta\omega_c(\phi)$, where $\phi = \tan^{-1}(y/x)$ and the spin is located on the central resonant surface. Note, that for given values of z and ϕ the coordinates y and x of the spin are fixed if the spin is located on the central resonant surface. The value of ϕ can be changed by rotating the cantilever about its axes. One can see the sharp extrema of the function $\delta\omega_c(\phi)$. Again, the small distance between the CT and the sample is preferable for the measurement of the "azimuthal position" of the spin.

Finally, we will consider the realistic case when the direction of polarization of the *rf* field $2\vec{B}_1$ is fixed in the laboratory system of coordinate. Now

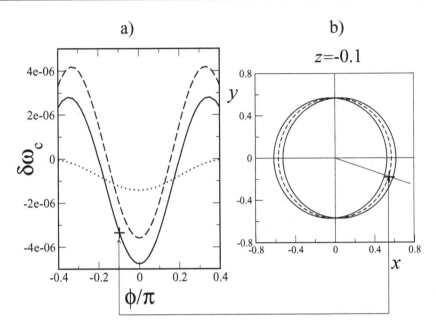

Figure 10.11: (a) Dependence $\delta\omega_c(\phi)$, with $\phi = \tan^{-1}(y/x)$ for the central resonant surface and $z = -0.1$ (full line); $z = -0.43$ (dashed line), $z = -1$ (dotted line). (b) solid line shows the cross-section of the resonant slice for $z = -0.1$. Dashed line shows the intersection between the plane $z = -0.1$ and the central resonant surface. The plus in (b) shows the spin location $\phi/\pi = -0.1$. The corresponding frequency shift is marked by a plus on (a).

the angle θ between the direction of polarization of $2\vec{B}_1$ and the field $\vec{B}_0$ depends on the spin coordinate as the magnitude and direction of the dipole field $\vec{B}_d^{(0)}$ depend on the spin location. To describe this case we ignore the component of $2\vec{B}_1$, which is parallel to $\vec{B}_0$, and change B_1 to $B_1 \sin\theta$ in all our formulas. As an example Fig. 10.12(a) demonstrates the dependence $\delta\omega_c(z)$ for the case when the rf field is polarized along the x-axis. One can see that in a narrow region of z, the magnitude of the frequency shift sharply drops.

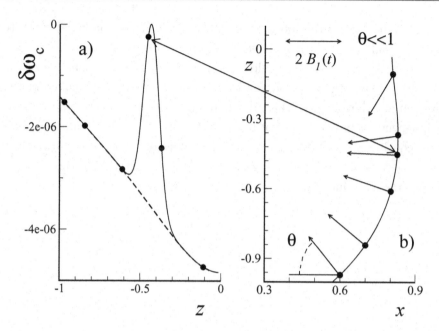

Figure 10.12: (a) Dependence $\delta\omega_c(z)$ when the rf field $\vec{B}_1$ is parallel to the x-axis. The spin is located at the central resonant surface $y = 0$, $x > 0$. Solid line are numerical data, dashed line is the analytical estimate (10.73), which assumes adiabatic motion of the spin magnetic moment $\vec{\mu}$ parallel to $\vec{B}_{eff}$. For a few numerical points indicated as black circles in (a) the corresponding $\vec{B}_0$ field is shown in (b). (b) Solid line: intersection between the central resonant surface and the $x - z$ plane. Arrows show the magnetic field $\vec{B}_0$ on this intersection at the points indicated as black circles in (a). The absolute value of the frequency shift $|\delta\omega_c|$ drops at the spin locations where $\vec{B}_0$ is approximately parallel to $\vec{B}_1$ ($\theta \ll 1$).

This occurs because in this region the magnetic field $\vec{B}_0$ is almost parallel to the x−axis (see Fig. 10.12(b)). Thus, the effective field $B_1 \sin\theta$ is small, the condition of the adiabatic motion (2.27), which in our case can be written as $\gamma[B_1 \sin\theta]^2 \gg |d\vec{B}_{eff}/dt|$, is not satisfied, and the spin does not follow the effective field. The dashed line in Fig. 10.12(a) corresponds to the analytical estimate (10.73) with the substitution $B_1 \to B_1 \sin\theta$: the analytical estimate assumes adiabatic conditions, which are violated for small θ.

The sharp drop of $|\delta\omega_c|$ could be observed either by the change of the distance between the CT and the sample surface or by the change of the direction of polarization of the *rf* field. In any case, this effect could be used for independent measurement of the spin "depth" below the sample surface.

In conclusion, we note that taking the values of parameters (10.79) and assuming $G_0 \approx G_z$ we obtain from Eq. (10.76) $\delta\omega_c/2\pi \approx 3.7$ mHz, which is close to the experimental value 4.2 mHz.

Chapter 11

Magnetic Noise and Spin Relaxation in OSCAR MRFM

In this chapter we will consider the direct interaction between the spin and its environment, which causes the magnetic noise on the spin. We will assume that the main source of magnetic noise on the spin is associated with the thermal vibrations of the CT. Magnetic noise causes the spin deviation from the effective field in OSCAR MRFM. This deviation, in turn, causes quantum jumps (spin flips). We have discussed quantum collapses and quantum jumps shortly in Chap. 8. In this chapter we will consider simple models for both phenomena (Berman *et al.* [33, 34]). We will start this chapter from the analysis of the OSCAR signal generated by an ensemble of spins rather than a single spin (Berman *et al.* [35–38]). In this case, the magnetic noise causes the decay of the regular OSCAR signal, which can be described as the spin relaxation in the RSC with the characteristic time τ_m. The regular OSCAR signal is followed by the random OSCAR signal, which does not decay with time like random quantum jump in the case of a single spin. We will discuss both regular and random OSCAR signals. We will also consider the opportunity of suppression of the magnetic noise using a nonuniform cantilever instead of a uniform one.

11.1   OSCAR relaxation in a spin ensemble

In this section we will consider the regular and random OSCAR signals generated by an ensemble of spins inside the resonant slice. First, we will consider the CT oscillating perpendicular to the sample surface. Figure 11.1 shows the resonant slice, where the Larmor frequencies ω_L will match the *rf* frequency ω in the process of the CT vibrations.

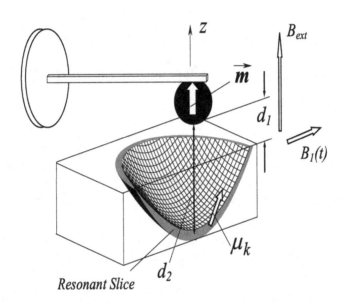

Figure 11.1: The resonant slice for the CT oscillating perpendicular to the sample surface. B_{ext} and $B_1(t)$ are the permanent and rotating *rf* magnetic fields; $\vec{m}$ is the magnetic moment of the ferromagnetic particle, $\vec{\mu}_k$ is the kth magnetic moment in the resonant slice.

The resonant Larmor frequency of a spin

$$\omega_L \approx \gamma(B_{ext} + B_{dz}),\qquad(11.1)$$

depends on the CT position z_c and on the position of the spin inside the resonant slice. The frequency ω_L for the spins on the upper boundary of the resonant slice matches the rf frequency ω when the CT is in its upper position $z_c = A$. For the same position of the CT the frequency ω_L for a spin on the bottom boundary of the resonant slice will be much smaller than ω. Note, that in Eq. (11.1) and below we assume that the external magnetic field B_{ext} is much greater than the dipole field B_d, produced by the CT.

We are not going to consider an analytical approach to the computation of the relaxation time τ_m used by Mozyrsky *et al.* [39], but restrict ourselves with analytical estimate and numerical simulations.

For analytical estimate we consider the motion of a single classical magnetic moment $\vec{\mu}_k$ at the center of the resonant slice, under the action of the fluctuating magnetic field produced by the ferromagnetic particle. Assume that the magnetic moment moves adiabatically, together with the effective field in the semi-plane $(+z) - (+x) - (-z)$, in the RSC. When the polar angle of the vector $\vec{\mu}_k$ is not small, the weak fluctuating dipole field produced by the ferromagnetic particle has a component perpendicular to $\vec{\mu}_k$ which causes a deviation of $\vec{\mu}_k$ from the effective field.

In the process of adiabatic reversals the resonance frequency of the magnetic moment in the RSC changes from ω_{max} (near the $+z-$axis) to ω_R (near the transverse plane), and back to ω_{max} (near the $-z-$axis). If a characteristic frequency of the fluctuating field falls in the region (ω_R, ω_{max}) the fluctuating field causes a noticeable deviation of the magnetic moment from the effective field.

Next, we make assumptions similar to that in Chap. 9, when we estimated the characteristic time interval between the quantum jumps. The modes in the narrow region, approximately $(\omega_R, 2\omega_R)$, have the greatest thermal amplitudes. The geometrical factor also favors these modes because the magnetic moment $\vec{\mu}_k$ has the resonant frequency in the region $(\omega_R, 2\omega_R)$ when it is very close to the transverse plane where the fluctuating field is perpendicular to the magnetic moment $\vec{\mu}_k$. To estimate the action of the fluctuating field, we will assume that in this region of the resonance frequencies the magnetic moment experiences the fluctuating magnetic field produced by the thermal CT vibrations. To simplify our estimate, we assume that these vibrations have the constant amplitude A_R^T, given by Eq. (9.14), and random phase. The characteristic time between the "phase jumps" of the fluctuating field (the correlation time) can be estimated as the Rabi period T_R. Assuming that the angular spin deviation from the effective field is a diffusion process we obtain the characteristic deviation during a single reversal $\langle \Delta \theta_1^2 \rangle$. The value of $\langle \Delta \theta_1^2 \rangle$ is given by Eqs. (9.20) where we should change $G_z = \partial B_z / \partial x$ to $G = \partial B_d / \partial z$. We may estimate the relaxation time τ_m as the time corresponding to the total deviation $\langle \Delta \theta^2 \rangle \simeq 1$. In this case τ_m is the time interval, which includes $1/\langle \Delta \theta_1^2 \rangle$ reversals

$$\tau_m = \frac{\cdot T_c}{2} \left(\frac{1}{\langle \Delta \theta_1^2 \rangle} \right). \tag{11.2}$$

Thus, we obtain the formula which is similar to (9.25) for the characteristic time interval between quantum jumps. Omitting the numerical coefficient we may rewrite this formula in the form convenient for the discussion:

$$\tau_m \sim \frac{k_c A}{\gamma G k_B T} \left(\frac{\omega_R}{\omega_c} \right)^2. \tag{11.3}$$

We shall discuss this formula. The characteristic time of the deviation from the effective field increases with increasing the Rabi frequency ω_R, but decreases with increasing temperature. These dependences reflect the obvious properties of the thermal noise of the cantilever. The dependence τ_m on the amplitude of the cantilever vibrations A is associated with the time Δt_1 of passing the resonant region $(\omega_R, 2\omega_R)$: the greater A is, the smaller the time Δt_1; thus the greater number of reversals is needed to provide a significant deviation of $\vec{\mu}_k$ from the effective field. The dependence τ_m on the magnetic field gradient G appears as the result of two competing factors. On the one hand, when the gradient G increases, the fluctuating magnetic field also increases. On the other hand, the time of passing the resonant region Δt_1 decreases when the gradient G increases.

Below, we describe the results of our computer simulations. We will consider the damped vibrations of the CT interacting with the spins of the resonant slice in the presence of the magnetic noise. The classical equation of motion for the CT is

$$\ddot{z}_c + z_c + \frac{1}{Q}\dot{z}_c = f(\tau),$$

$$f(\tau) = \sum_k \eta_k \, \mu_{kz}, \tag{11.4}$$

$$\eta_k = \frac{\mu_0}{4\pi} \frac{3m\mu}{k_c A^5} \frac{\tilde{z}_k \left(5\tilde{z}_k^2 - 3\tilde{r}_k^2\right)}{\tilde{r}_k^7}.$$

Here we use the dimensionless time $\tau = \omega_c t$, the CT coordinate z_c is written in units of the CT amplitude A, x_k, y_k and z_k are the coordinates of the kth magnetic moment $\vec{\mu}_k$ in the same units,

$$\tilde{z}_k = z_k - z_c, \qquad \tilde{r}_k = \left(x_k^2 + y_k^2 + \tilde{z}_k^2\right)^{1/2}, \tag{11.5}$$

the magnetic moment $\vec{\mu}_k$ is written in units of its magnitude μ. In this section we will consider $\vec{\mu}_k$ not as a magnetic moment of the kth spin with magnitude μ_B but as a macroscopic magnetic moment of an arbitrary magnitude μ, which is the same for all k. In our simulations the value of μ is adjusted to the number of magnetic moments N, so that the average magnetization M in the resonant slice remains constant.

The motion of the kth magnetic moment in RSC is given by

$$\dot{\mu}_{kx} = -\Delta_k \, \mu_{ky},$$

$$\dot{\mu}_{ky} = \Delta_k \, \mu_{kx} - \epsilon \, \mu_{kz}, \tag{11.6}$$

$$\dot{\mu}_{kz} = \epsilon \, \mu_{ky}.$$

In Eqs. (11.6), the following notation were used:

$$\Delta_k = \frac{(\gamma B_{ext} - \omega)}{\omega_c} + \frac{\mu_0}{4\pi} \frac{\gamma m}{\omega_c A^3} \frac{3\tilde{z}_k^2 - \tilde{r}_k^2}{\tilde{r}_k^5},$$

$$\tag{11.7}$$

$$\epsilon = \frac{\omega_R}{\omega_c}.$$

Note, that Δ_k is the z-component of the rotating-frame effective field $\vec{B}_{eff}^k$ (in units ω_c/γ), and ϵ is the x-component of the effective field in the same units.

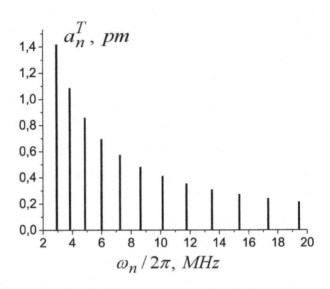

Figure 11.2: The thermal amplitudes of the high frequency cantilever modes for a silicon cantilever (190 μm $\times$ 3 μm $\times$ 850 nm) used in [26]. The cantilever temperature is 80 K.

The upper and the lower boundaries of the resonant slice are determined by the equation

$$\Delta_k = 0, \qquad \text{at} \qquad z_c = \pm 1. \tag{11.8}$$

In our numerical experiments, the magnetic moments have been distributed uniformly inside the resonant slice. We used the following initial conditions,

$$z_c = -1, \qquad \mu_{kz} = 1, \tag{11.9}$$

i.e. the magnetic moments are oriented approximately along the effective field in the rotating frame. To model the feedback technique in OSCAR MRFM, our computer algorithm increased the value of z_c to 1 every time the cantilever passed the upper point. The period of the cantilever oscillations was determined as the time interval between the instants of the z_c maximum values.

To model the magnetic noise on spins we use the following approach. First, we estimate the thermal amplitude a_n^T of the nth cantilever mode using the equipartition theorem:

$$\frac{1}{2} m_c \left(\omega_n a_n^T \right)^2 = k_B T. \tag{11.10}$$

As an example, Fig. 11.2, demonstrates the thermal amplitudes of the high frequency cantilever modes for the experiments of Stipe et al. [26].

As we mentioned in Chap. 4 the cantilever mass m_c is connected to the effective mass m^* by the relation $m^* = m_c/4$. Also, the amplitude of the CT vibrations for any mode is twice the amplitude of the mode. Thus, for the nth cantilever mode the thermal amplitude A_n^T of the CT vibrations can be written as

$$A_n^T = 2a_n^T = \frac{\omega_c}{\omega_n} \left(\frac{2k_B T}{k_c} \right)^{1/2}. \tag{11.11}$$

To describe the influence of the noise on the spin dynamics we replace the coordinate z_c with $z_c + \delta z_c$ in the expressions for $\tilde{z}_k$ and $\tilde{r}_k$ in Eq. (11.7), where

$$\delta z_c = \sum_n \frac{A_n^T}{A} \cos(\Omega_n \tau + \psi_n),$$

$$\Omega_n = \frac{\omega_n}{\omega_c}, \tag{11.12}$$

and ψ_n is a random phase. We solve the system of Eqs. (11.4) and (11.6) changing $z_c \rightarrow z_c + \delta z_c$ according to Eq. (11.12). We did not take into account the influence of the thermal noise on η_k in Eqs. (11.4). As our simulations demonstrate, the influence of the thermal noise does not cause a significant direct contribution to the CT vibrations through the parameter η_k, but causes a dephasing of the magnetic moments in the resonant slice. The number of magnetic moments in the resonant slice was 200 ($1 \leq k \leq 200$). Our simulations show that the results do not change significantly when this number is increased to 400.

The parameters in our numerical simulations were taken from experiment [26]:

$$B_{ext} = 140 \text{ mT}, \quad k_c = 14 \text{ mN/m},$$
$$\omega_c/2\pi = 21.4 \text{ kHz}, \quad A = 28 \text{ nm}, \quad (11.13)$$
$$Q = 2 \times 10^4, \quad m = 1.5 \text{ pJ/T}.$$

The distance between the bottom of the ferromagnetic particle and the surface of the sample, d_1 (see Fig. 11.1), was taken $d_1 = 700$ nm, and the radius of the ferromagnetic particle was equal to d_1. The average magnetization of the sample was $M = 0.89$ A/m. For these values of parameters the value of d_2 in Fig. 11.1 was found to be $d_2 = 875$ nm, and the value of the magnetic field gradient at the center of the resonant slice was 140 kT/m.

We have studied the decay of the OSCAR signal, $\Delta T_c/\Delta T_0$, where ΔT_c is the shift of the period of the CT vibrations due to the influence of the magnetic moments of the sample, and ΔT_0 is the initial value of ΔT_c. We express the amplitude A_n^T in terms of the amplitude A_R^T of the thermal vibrations near the Rabi frequency. From Eq. (9.14) for A_R^T we obtain the obvious relation $A_n^T = A_R^T \omega_R/\omega_n$. The phases ψ_n in Eq. (11.12) were changed randomly between 0 and 2π with random time intervals, τ_ψ, between two successive changes of the phase. In particular, we put $\tau_\psi = N'\tau_0$, where τ_0 takes random values between $2\pi/1.2\epsilon$ and $2\pi/0.8\epsilon$, and N' is a free parameter of the model.

Our simulations show that the decay of the OSCAR signal is almost independent of N' for $N' < 1000$. (See Fig. 11.3.) We have found that for $N' < 1000$ the signal can be approximately described by an exponential function with relaxation time, τ_m. (See Fig. 11.4.)

To study the spin relaxation we also considered values of parameters which provide relatively small relaxation time. This allowed us to reduce the computation time and to determine the characteristic scaling properties of the relaxation process. As an example, Fig. 11.4 shows the decay of the OSCAR signal for various values of the CT amplitudes. One can see that the relaxation time τ_m decreases when the CT amplitude decreases. Note

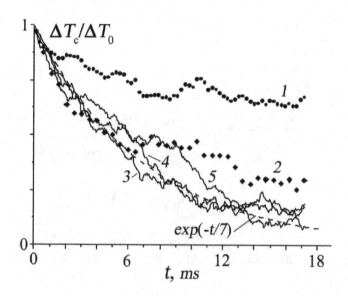

Figure 11.3: Decay of the OSCAR signal for different values of the parameter N'. Curves $1 - 5$ corresponds to $N' = 10^5, 10^3, 100, 10$ and 2. The amplitudes are $A_R^T = 5$ pm, $A = 15$ nm, and $\epsilon = 390$.

that if we take into consideration spins near the resonant slice, the value of τ_m slightly increases. (Compare curves b and c in Fig. 11.4.)

Figure 11.5 shows the decay of the OSCAR signal for various values of the temperature and the *rf* field B_d. One can observe the expected decrease of the relaxation time τ_m with an increase of the temperature or decrease of the rotating magnetic field amplitude.

Figure 11.6 shows the decay of the OSCAR signal for various numbers of high-frequency modes taken into considerations. One can see that the relaxation time τ_m slightly decreases when the number of high-frequency cantilever modes increases from 2 to 22. A further increase of the number of modes does not essentially change the decay rate.

Next, we will consider the random OSCAR MRFM signal, which follows the decay of the regular one. In Figs.11.7–11.9 we took into consideration 22 high frequency modes. The lowest mode had the frequency closest to the

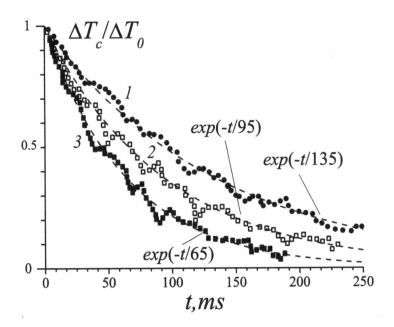

Figure 11.4: Decay of the OSCAR signal for various values of the CT amplitude A; $A_R^T = 1$ pm and $\epsilon = 390$. Curve 1 corresponds to $A_0 = 15$ nm; curves 2 and 3 correspond to $A_0 = 7.5$ nm. For curve 2, we also took into consideration spins close to the resonant slice.

Rabi frequency ω_R. The characteristic time interval between the phase jumps was taken as 10 Rabi periods ($N' = 10$).

Figure 11.7 demonstrates the connection between the regular and random OSCAR MRFM signals, for a system of 50 magnetic moments uniformly distributed in the resonant slice. The decay time of the regular signal is $\tau_m = 145$ ms. The MRFM random signal is about 30% of the maximum regular signal.

Figure 11.8 shows the standard deviation of the random signal

$$\sigma_\xi = \sqrt{\langle (\xi - \langle \xi \rangle)^2 \rangle},$$

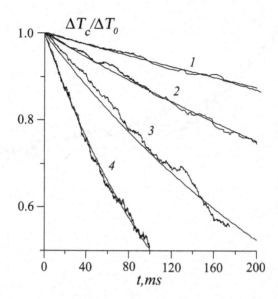

Figure 11.5: Decay of the OSCAR signal for $A = 28$ nm and various values of the temperature T and the rf field B_1. Curves 1-3 corresponds to $\epsilon = 390$ ($B_1 = 0.3$ mT) and $T = 20$ K, 40 K and 80 K. Curve 4 corresponds to $\epsilon = 195$ ($B_1 = 0.15$ mT) and $T = 80$ K. The relaxation times, τ_m, for curves $1-4$ are 1500 ms, 700 ms, 310 ms and 145 ms.

$$\xi = \frac{\Delta T_c}{\Delta T}, \tag{11.14}$$

as a function of a number of magnetic moments distributed in the resonant slice (for the same value of the average magnetization M). One can see that with the increase of N the standard deviation σ_ξ approaches the value 0.1. This indicates that the random MRFM signal survives the transition to the continuous magnetization.

As an illustration to the spin dynamics, Fig. 11.9 shows the random change of the magnetic moment component along the effective field. The magnetic moment was arbitrary chosen in the resonant slice. One can see that the magnetic moment randomly moves between the direction of the effective field and the opposite direction.

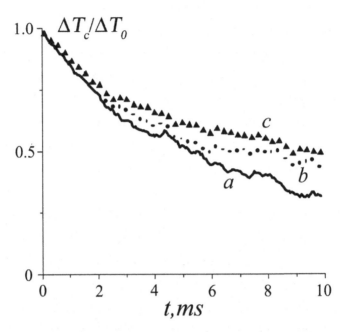

Figure 11.6: Decay of the OSCAR signal for $A_R^T = 5$ pm and $A = 15$ nm, and various numbers of the high-frequency modes taken into considerations. The lowest of high-frequency cantilever modes is the mode with the frequency closest to the Rabi frequency $\omega_R = 8.4$ MHz ($\epsilon = 390$). Curves a, b and c correspond to $22, 3$ and 2 high-frequency cantilever modes, respectively, including the lowest one.

Next, we will describe the results of our computer simulations for the CT oscillating parallel to the sample surface. Figure 11.10 shows the resonant slice, or better to say, semi-slice for this case. Again, the magnetic moments $\vec{\mu}_k$ of the same magnitude μ are distributes inside the resonant slice, and the external magnetic field B_{ext} is assumed to be much greater than the dipole field $\vec{B}_d$, produced by the CT on the spins. Again, the value of μ is adjusted to the number N of magnetic moments in the slice, so that the magnetization M remains constant.

The equations of motion for the CT and the magnetic moments are similar

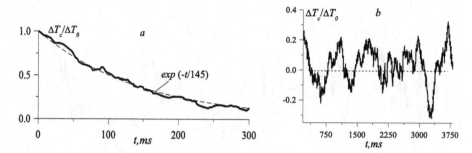

Figure 11.7: (a) The regular MRFM signal. (b) The random MRFM signal. The temperature is 80 K. The Rabi frequency $\omega_R/2\pi = 4.17$ MHz. The cantilever amplitude $A = 28$ nm.

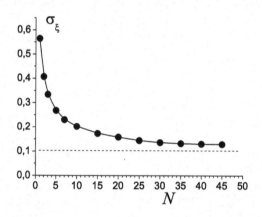

Figure 11.8: The standard deviation of the random MRFM signal as a function of the number N of magnetic moments in the resonant slice (at a fixed value of the average magnetization $M = 0.89$ A/m). All parameters are the same as in Fig. 11.7.

to Eqs. (11.4) and (11.6). For the CT coordinate we have:

$$\ddot{x}_c + x_c + \frac{1}{Q}\dot{x}_c = f(\tau),$$

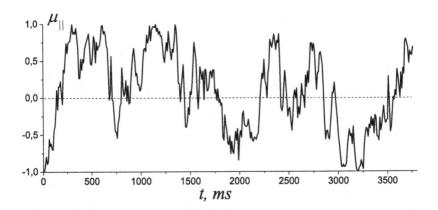

Figure 11.9: The component $\mu_{\parallel}$ of an arbitrary chosen magnetic moment along the effective field in the RSC. All parameters are the same as in Fig. 11.7.

$$f(\tau) = \sum_k \eta_k \mu_{kz},$$
(11.15)

$$\eta_k = \frac{\mu_0}{4\pi} \frac{3m\mu}{k_c A^5} \frac{\tilde{x}_k \left(5z_k^2 - 3\tilde{r}_k^2\right)}{\tilde{r}_k^7},$$

$$\tilde{x}_k = x_k - x_c, \qquad \tilde{r}_k = (\tilde{x}_k^2 + y_k^2 + z_k^2)^{1/2}.$$

For magnetic moments we have again Eqs. (11.6). In (11.7), we change $\tilde{z}_k \to z_k$. Instead of Eq. (11.12) we have the similar equation for δx_c:

$$\delta x_c = \sum_n \frac{A_n^T}{A} \cos\left(\Omega_n \tau + \psi_n\right).$$
(11.16)

Below we show the results of our computer simulations for the following values of parameters:

$$\omega_c/2\pi = 7 \text{ kHz}, \quad k_c = 100\mu \text{ N/m}, \quad Q = 5 \times 10^4,$$

$$A = 10 \text{ nm}, \quad m = 25 \, f\text{J/T}, \quad d_1 = 220 \text{ nm}, \quad d_2 = 300 \text{ nm},$$

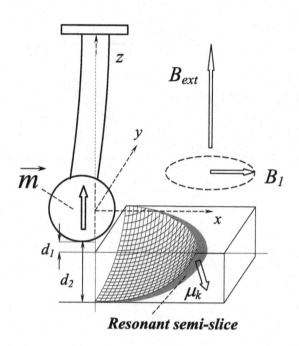

Resonant semi-slice

Figure 11.10: Resonant semi-slice for the CT oscillating parallel to the sample surface. The $z - y$ plane is the plane of symmetry for the resonant slice.

$$B_1 = 300 \ \mu\text{T}, \quad \omega/2\pi = 3 \text{ GHz}, \quad M = 0.9 \text{ A/m}, \tag{11.17}$$

and the radius of the ferromagnetic particle 200 nm.

The resonant slice boundaries were found from the condition $\Delta_k = 0$ for $x_c = \pm 1$. We took into account only moments $\vec{\mu}_k$ located in the upper part of the resonant slice (see Fig. 11.11). They provide the main contribution to the MRFM signal.

Note, that for two spins with coordinates (x, y, z) and $(-x, y, z)$, the change of the z-component of the dipole field, caused by the cantilever displacement, has opposite signs at two locations. Let, for example, the z-component of the effective field for this two spins be zero when the ferro-

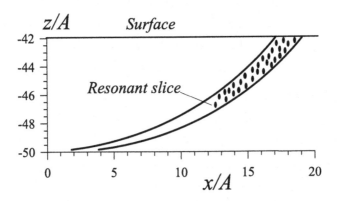

Figure 11.11: The cross-sectional area of the resonant semi-slice in the $x - z$ plane. Dots show random distribution of magnetic moments in the upper part of the resonant slice.

magnetic particle is at the origin ($x_c = 0$). If $x_c \neq 0$, the z-component of the effective field on the first spin is opposite to that on the second spin. If the initial direction of two spins relative to the external magnetic field is the same, these spins will have an opposite direction relative to the local effective field. Thus, the two spins induce a MRFM signal of the opposite sign. If the spins are uniformly distributed in the resonant slice, the net MRFM signal disappears. That is why, in our simulations, we assume that spins occupy only the resonant semi-slice $x_c > 0$. The cross-sectional area of the resonant semi-slice in the $x - z$ plane is shown in Fig. 11.11.

Note, that only initial decaying MRFM signal disappears for uniformly distributed spins. The random MRFM signal, which follows the initial regular signal, would have been present if we consider the whole resonant slice.

We assume that initially all magnetic moments point in the positive z direction, and the CT is in its right end point, $x_c = 1$. In our simulations the phases ψ_n in Eq. (11.16) were changed randomly between 0 and 2π with random time intervals between two successive "jumps". The time inter-

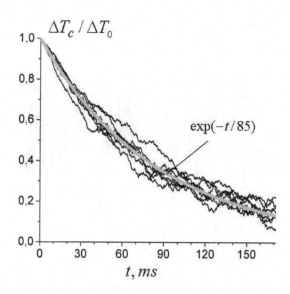

Figure 11.12: The decay of the regular MRFM signal for seven noise realizations. The temperature is 20 K. The number of magnetic moments in the resonant semi-slice is $N = 100$.

val between the phase jumps was taken randomly between $8.3T_R$ and $14T_R$, where T_R is the Rabi period. In Eq. (11.16), we took into consideration 25 cantilever modes in the vicinity of the Rabi frequency. The regular MRFM signal decays approximately exponentially (see Fig. 11.12).

The regular MRFM signal is followed by the non-decaying random signal. Figures 11.13–11.15 illustrate the results of computer simulations for the random signal. Figure 11.13 shows the random signal which follows the regular signal presented in Fig. 11.12. The spectrum of the random signal does not have a unique characteristic form. At the same time, the amplitude of harmonics A_m in the low-frequency region with the upper boundary of the order $1/(4\tau_m)$ is higher than in the high-frequency region of the spectrum. (See Fig. 11.14.)

The characteristic decay time τ_m of the regular signal can be estimated as a

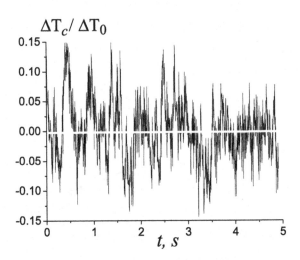

Figure 11.13: The random MRFM signal which follows the regular signal shown in Fig. 11.12.

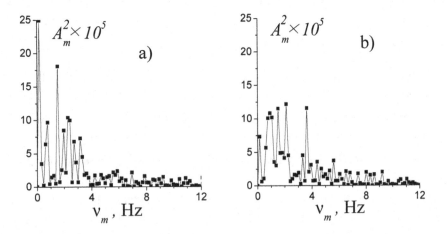

Figure 11.14: Characteristic random signal spectra for two different realizations: $\Delta T_c(t)/\Delta T_0 = \sum_m A_m \cos(2\pi\nu_m t + \psi_m)$. A value on the vertical axis, e.g. 5, corresponds to $A_m^2 = 5 \times 10^{-5}$.

quarter of period of the boundary frequency in the random signal spectrum.

Finally, Fig. 11.15 shows the random signal spectral density $\langle A_m^2(\nu_m)\rangle$ averaged over seven realizations of the noise. Note that the amplitudes for the lowest frequencies are computed with an error because their periods are compared with the total time of simulations of the signal.

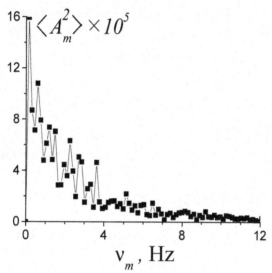

Figure 11.15: The noise spectral density of the random signal averaged over seven realizations of the noise .

11.2   Reduction of magnetic noise

The reduction of magnetic noise on the spin is extremely important for any MRFM technique. Such a reduction would increase the time interval t_{jump} between the quantum jumps in a single spin MRFM and relaxation time τ_m for an ensemble of spins. To reduce the magnetic noise one should suppress the thermal vibrations of the CT near the Rabi frequency. This can be achieved using a nonuniform (loaded) cantilever instead of a uniform one.

In this section we consider two ways for the reduction of the magnetic noise using a cantilever with a nonuniform thickness. The first one is to reduce the values of the cantilever eigenfunctions near the CT for the frequencies, which are close to the Rabi frequency. Better to say, one should reduce the values of the cantilever eigenfunctions near the location of the ferromagnetic particle which, in general, can be shifted from the CT. The second way, suggested by Chui *et al.* [40] is to increase the gap between the eigenfrequencies of the cantilever in the region of the Rabi frequency.

For a cantilever with a nonuniform thickness, instead of Eq. (4.25), we have the following equation of motion:

$$
\rho \, S(x) \frac{\partial^2 z_p}{\partial t^2} = -Y \frac{\partial^2}{\partial x^2} \left[I(x) \frac{\partial^2 z_p}{\partial x^2} \right], \tag{11.18}
$$

with the boundary conditions (4.26). Instead of Eq. (4.27), we have the following equation for the eigenfunctions $f_j'(x)$ and eigenfrequencies ω_j':

$$
\rho \, S(x) \, f_j'(x)(\omega_j')^2 = Y \frac{\partial^2}{\partial x^2} \left[I(x) \frac{\partial^2}{\partial x^2} f_j'(x) \right]. \tag{11.19}
$$

The eigenfunctions $f_j'(x)$ can be normalized to the cantilever volume V_c:

$$
\int_0^{l_c} dx \, S(x) f_j'(x) f_m'(x) = \delta_{jm} V_c, \quad V_c = \int_0^{l_c} dx \, S(x). \tag{11.20}
$$

Note, that the condition of orthogonality for $j \neq m$ includes the factor $S(x)$.

To find the eigenfunctions $f_j'(x)$ of the nonuniform cantilever we use the expansion over the eigenfunction $f_j(x)$ of the uniform cantilever

$$
f_j'(x) = \sum_{k=1}^{M} c_{kj} f_k(x). \tag{11.21}
$$

We take the number of basis functions $M = 50$ which provides the accurate approximation for the eigenfunctions of the nonuniform cantilever. In particular, we have verified the conditions of orthogonality (11.20) for the normalized eigenfunctions $f_j'(x)$: the value of the integral in the left-hand side of (11.20) did not exceed 0.02.

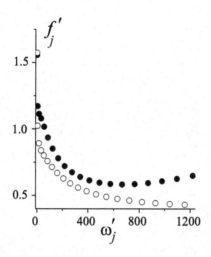

Figure 11.16: Dependence of the values of eigenfunctions $f'_j(l_c)$ on the eigen-frequencies ω'_j. Dots corresponds to $\gamma = 4, x_0 = 0.095, \delta = 0.05$. The mass of the cantilever (in terms of the mass of the "unperturbed" uniform cantilever) is $m_c = 1.38$. The fundamental frequency (in terms of the "unperturbed" frequency ω_c) is $\omega_1 = 0.65$. Circles correspond to $\gamma = 6$, $x_0 = 0.99$, $\delta = 0.05$, $m_c = 1.325$, $\omega_1 = 0.68$. For $x > x_0$ the cantilever thickness is constant.

In our computations for a given value of j, we have found the eigenvectors c_{kj} ($1 \le k \le M$) of the matrix α_{km},

$$l_c(\omega'_j)^2 c_{kj} = \sum_{m=1}^{M} \alpha_{km} \, c_{mj},$$

$$(11.22)$$

$$\alpha_{km} = \frac{Y}{\rho} \int_0^{l_c} dx \left[\frac{\partial^2}{\partial x^2} \left(\frac{f_k(x)}{S(x)} \right) \right] I(x) \frac{\partial^2 f_m(x)}{\partial x^2}.$$

We have considered the uniform increase of the thickness of the cantilever

$$t'_c(x) = t_c\{1 + \gamma \exp[-(x - x_0)^2/\delta^2]\},$$

$$(11.23)$$

where t_c is the thickness of the "unperturbed" cantilever, $(1 + \gamma)t_c$ is the

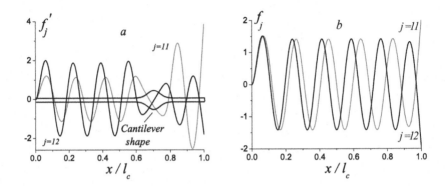

Figure 11.17: (a) Eigenfunctions $f_j'(x)$ for $j = 11$ and $j = 12$ (the corresponding eigenfrequencies $\omega_{11}' = 362$ and $\omega_{12}' = 416$). The values of parameters: $\gamma = 4$, $x_0 = 0.7$, $\delta = 0.05$, $\omega_1 = 0.82$. (b) Eigenfunctions $f_j(x)$ for the "unperturbed" uniform cantilever.

maximum thickness which is achieved at the point $x = x_0$, and δ is the characteristic size of inhomogeneity.

For $t_c \ll w_c \ll l_c$ the eigenfrequencies and the eigenfunctions depend on the ratio l_c/t_c which was chosen as $95/3$ in our simulations. In some cases, for $x > x_0$ we take the constant value of the thickness : $t_c'(x) = t_c(1 + \gamma)$ (see captions to Figs. 11.16, 11.19 and 11.20).

Typically, a ferromagnetic particle in MRFM experiments is placed at the CT. (Except for this section we always consider only such case.) Our simulations show that in this case, one should increase the thickness of the cantilever near its tip in order to provide the maximum possible reduction of the cantilever eigenfunction values near the tip.

Figure 11.16 shows the values of the eigenfunctions $f_j'(x)$ near the tip $(x = l_c)$ as a function of the eigenfrequency ω_j'. In Fig. 11.16 and below the ratio $l_c/t_c = 190/6$, the eigenfrequencies of the modes ω_j' are given in units of the fundamental frequency ω_c of the unperturbed uniform cantilever. The values x_0 and δ are given in units of l_c.

We have found that even a more significant effect in the noise reduction

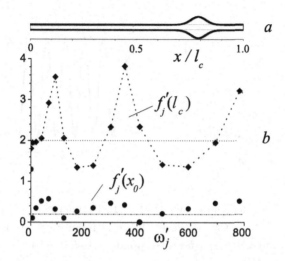

Figure 11.18: (a) The cantilever shape; (b) the values of the eigenfunctions at the center of the inhomogeneity $x = x_0$ (circles) and near the cantilever tip $x = l_c$ (squares). The values of parameters: $\gamma = 3, x_0 = 0.79, \delta = 0.05, \omega_1' = 0.8$.

can be achieved if the ferromagnetic particle is located at a some distance from the CT. In this case, the increase of the cantilever thickness should be centered at the position of the ferromagnetic particle. Figure 11.17 demonstrates the reduction of the eigenfunctions f_{11}' and f_{12}' in the region of the inhomogeneity. Circles in Fig. 11.18(b) show the value of the eigenfunctions at the center of the inhomogeneity $x = x_0 < l_c$. One can see that the values of $f_j'(x_0)$ may become very small. At the same time, the values $f_j'(l_c)$ increase in comparison with the uniform cantilever (squares in Fig. 11.18).

Figure 11.19 demonstrates the same features as Fig. 11.18 for a cantilever of a different shape. One can see that in the frequency region ω_{eff} the values $f_j'(x_0)$ are very small: $|f_j'(x_0)| \ll 1$.

For a cantilever with a mass increase about 100%, the reduction of values of the eigenfunctions is not significant. However, the increase of the gap between the eigenfrequencies is rather large. Figure 11.20 demonstrates both the change of the values of eigenfunctions near the tip and the "repulsion"

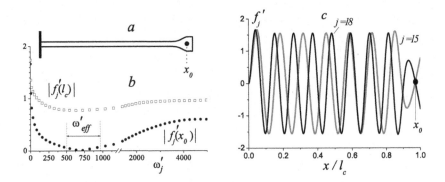

Figure 11.19: (a) and (b) The same as in Fig. 11.18 but for $\gamma = 2$, $x_0 = 0.97$, $\delta = 0.05$, $\omega_1' = 0.8$. For $x > x_0$ the cantilever thickness is constant. (c) Eigenfunctions $f_j'(x/l_c)$ for $n = 15$ and $n = 18$ (the corresponding eigenfrequencies are $\omega_{15}' = 630$ and $\omega_{18}' = 924$).

of the eigenfrequencies, for such a cantilever. Using the method described in the previous section we computed the decay of the MRFM signal $\Delta T_c/\Delta T_0$. Figure 11.21 shows the typical decay of the MRFM signal for the following parameters: the amplitude of the CT vibrations is 15 nm, the effective spring constant and the fundamental frequency of the "unperturbed" uniform cantilever are 14 mN/m and 21 kHz, correspondingly, the Rabi frequency is 8.2 MHz, the magnetic moment of the ferromagnetic particle is $1.5 \times$ pJ/T, the radius of the ferromagnetic particle is 700 nm, the distance from the bottom of the ferromagnetic particle to the center of the resonant slice is 875 nm, the cantilever quality factor is 2×10^4. The cantilever oscillates perpendicular to the surface of the sample. To reduce the computational time we took the room temperature.

Curve (a) in Fig. 11.21 corresponds to the uniform cantilever with the ferromagnetic particle at the CT, curve (b) corresponds to the cantilever with the ferromagnetic particle and the inhomogeneity at the tip (dots in Fig. 11.16 show the eigenfunctions for this cantilever), curves (c) and (d) correspond to the cantilever with the ferromagnetic particle placed at some distance from

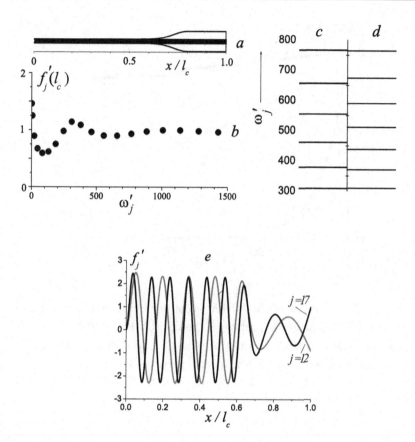

Figure 11.20: The values of eigenfunctions near the CT and the eigenfrequencies for a cantilever with the mass increase about 100%; (a) the cantilever shape; (b) the values of eigenfunctions near the tip; (c) the eigenfrequencies of the cantilever; (d) the eigenfrequencies of the "unperturbed" uniform cantilever; (e) eigenfunctions $f'_j(x)$ for $j = 12$ and $j = 17$ (the corresponding eigenfrequencies $\omega_{12} = 555$ and $\omega_{17} = 1138$). The values of parameters: $\gamma = 4$, $x_0 = 0.8$, $\delta = 0.1$, $m_c = 2.15$, $\omega_1 = 0.5$. For $x > x_0$ the cantilever thickness is constant.

the tip: $x = x_0 < l_c$, (curve (c) is for the cantilever shown in Fig. 11.19). The ratio of the relaxation times for these four cases is $1 : 10 : 63 : 140$.

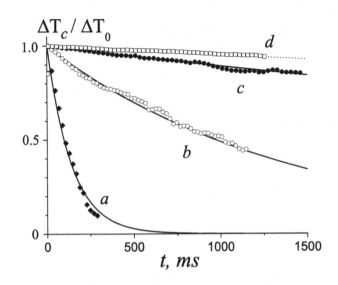

Figure 11.21: Typical decay of the MRFM signal for four cases: (a) the uniform cantilever with the ferromagnetic particle at the tip; (b) the nonuniform cantilever with the ferromagnetic particle at the tip ($\gamma = 4$, $x_0 = 0.95$, $\delta = 0.05$); (c) the nonuniform cantilever with the ferromagnetic particle at $x = x_0 < l_c$ ($\gamma = 3$, $x_0 = 0.79$, $\delta = 0.05$); (d) the same for $\gamma = 2$, $x_0 = 0.97$, $\delta = 0.05$.

For a cantilever with the mass increase about 100% shown in Fig. 11.20 with the ferromagnetic particle placed at the CT ($x = l_c$), the decay of the MRFM signal is almost the same as that described by the curve (d) in Fig. 11.21.

Thus, one can see that a small inhomogeneity of the cantilever thickness (with the mass change less than 50%) near the CT can provide a tenfold increase of the MRFM relaxation time. A greater suppression of the magnetic noise can be achieved if the ferromagnetic particle is located at some distance from the CT. The same results can be obtained using a cantilever with a mass increase about 100%.

11.3   Simple model for quantum jumps

Now we return to a single spin detection in OSCAR MRFM and consider a simple model, which describes the statistical characteristics of quantum jumps. We will consider a vertical cantilever with the CT oscillating along the x-axis, which is parallel to the surface of the sample (see Fig. 9.1). Let us consider the quantum Hamiltonian (8.4) with $z_c \to x_c$, $\dot{\phi} = 0$, and add the term, which describes the magnetic noise on the spin. We have

$$\mathcal{H} = \frac{1}{2}\left(\hat{p}_c^2 + x_c^2\right) + \epsilon\hat{S}_x - 2\eta\hat{S}_z x_c + \Delta(\tau)\hat{S}_z,$$

(11.24)

$$\Delta(\tau) = \frac{\gamma}{\omega_c}\Delta B_z(\tau).$$

We use the same notation and quantum units as in Chaps. 7 and 9.

We will consider a simple case when the external magnetic field is much greater than the dipole field on the spin. In this case the parameter η of the CT-spin interaction can be expressed in terms of $G_z = \partial B_z/\partial x$:

$$\eta = \frac{\mu_B G_z}{(\hbar\omega_c k_c)^{1/2}}.$$

(11.25)

The function $\Delta(\tau)$ in Eq. (11.24) is proportional to the random magnetic field $\Delta B_z(\tau)$ acting on the spin. We consider only the z-component of the random magnetic field, which is the most significant for quantum jumps. Indeed, the z-component of the random field causes the spin deviation from the effective field in the most "vulnerable" region near the $x - y$ plane, where the resonant frequency in the RSC is close to its minimum value ω_R. We will use the same parameters as in Sec. 1 of Chap. 8. For these parameters we have $\eta = 0.078$, $\epsilon = 1270$.

Below we describe our simplified model. We consider the function $\Delta(\tau)$ to be a random telegraph signal with two values $\pm\Delta$. The value of Δ can be expressed in terms of the amplitude A_R^T of the thermal CT vibrations near the Rabi frequency:

$$\Delta = 2\eta A_R^T.$$

(11.26)

We take the time interval between two consecutive "kicks" of $\Delta(\tau)$ randomly from the interval $(\tau_0 - \delta\tau, \tau_0 + \delta\tau)$, with the average time interval, τ_0, close to the Rabi period τ_R:

$$\tau_R = \frac{2\pi}{\epsilon} = 4.95 \times 10^{-3}. \tag{11.27}$$

We assume that every "kick" provided by the function $\Delta(\tau)$ is followed by the collapse of the wave function. Before the kick, the spin points in (or opposite to) the direction of the effective field. After the kick there appears the finite angle $\Delta\Theta$ between the new direction of the effective field and the average spin. Let, for example, a kick occurs at $\tau = \tau_k$ and, before the kick at $\tau = \tau_k - 0$ the spin points in the direction of the effective field $\vec{B}_{eff}(\tau_k - 0) = [\epsilon, 0, -2\eta x_c(\tau_k) + \Delta(\tau_k - 0)]$. The direction (the polar angle) of the spin Θ_{spin} and of the effective field Θ_{eff} are the same:

$$\Theta_{spin} = \Theta_{eff} = \tan^{-1}\left(\frac{B^x_{eff}}{B^z_{eff}(\tau_k - 0)}\right). \tag{11.28}$$

After the kick, the direction of the effective field Θ'_{eff} is

$$\Theta'_{eff} = \tan^{-1}\left(\frac{B^x_{eff}}{B^z_{eff}(\tau_k + 0)}\right). \tag{11.29}$$

The value of $\Delta\Theta$ is given by:

$$\Delta\Theta = \Theta_{spin} - \Theta'_{eff}. \tag{11.30}$$

As we explained in Chap. 8 (Sec. 1), the probability for the spin to "accept" the "before-kick" direction relative to the new effective field is $\cos^2(\Delta\Theta/2)$. The probability to "accept" the opposite direction, i.e. the probability of a quantum jump is $\sin^2(\Delta\Theta/2)$. A significant probability of a quantum jump appears only when the effective field passes the transversal $x - y$ plane. Thus, after every kick of the random field our computer code decides the "fate" of the spin in accordance with the probabilities of two events: to restore the previous direction relative to the effective field, or

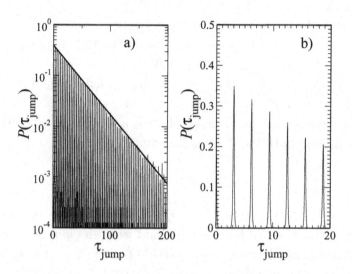

Figure 11.22: (a) Distribution function of time intervals between two consecutive quantum jumps for $\Delta = 100$, $\tau_0 = 0.01$ and 10^9 kicks; the solid line is a fit with $\tau_d = 32$. (b) Enlargement of (a)

to experience a quantum jump. In our model the CT experiences harmonic oscillations

$$x_c^{(\pm)}(\tau) = A\cos(1 \pm \delta\omega_0)\tau, \tag{11.31}$$

where ($\pm$) correspond to two CT trajectories with the spin pointing in (or opposite to) the direction of the corresponding effective field, and $\delta\omega_0$ is taken in units of ω_c. The dimensionless CT amplitude for our parameters is $A = 1.2 \times 10^5$, and the CT frequency shift estimated from Eq. 10.76 is

$$\delta\omega_0 = \frac{2\mu_B G_z}{\pi k_c A} = 4.2 \times 10^{-7}. \tag{11.32}$$

Note, that our model contains two important simplifications: first, we assume that the wave function collapse occurs immediately after the "kick" of the random field. Thus, we ignore the finite time when the spin-CT system is in an entangled state. Second, in a real situation the deviation of the spin from the effective field is a "quasi–resonance" process caused by the cantilever modes whose frequencies are close to the Rabi frequency. In our model this deviation appears as a result of the "kick" of the random field. Below we describe the results of our computer simulations. Figure 11.22 demonstrates a typical distribution of time intervals, τ_{jump}, between two consecutive quantum jumps. The distribution is a sequence of sharp peaks at $\tau_{jump} = \tau_n = n\pi$ with the Poisson-like amplitude

$$P(\tau_n) \sim \exp(-\tau_n/\tau_d). \tag{11.33}$$

Certainly, $P(\tau_{jump}) = 0$ at $\tau < \tau_0 - \delta\tau$. The sharp peaks appear as the probability of the quantum jump is significant when the spin passes through the transversal plane, i.e. every half-period of the CT oscillation which is equal to π. The average value of the time interval $\langle\tau_{jump}\rangle$ was found to be

$$\langle\tau_{jump}\rangle \approx \tau_d, \tag{11.34}$$

with an error less than 3%. The standard deviation is equal to τ_d with the same accuracy

$$(\langle\tau_{jump}^2\rangle - \langle\tau_{jump}\rangle^2)^{1/2} \approx \tau_d. \tag{11.35}$$

We studied the dependence of the average value $\langle\tau_{jump}\rangle$ on the parameters of our model. We have found that $\langle\tau_{jump}\rangle$ does not depend on $\delta\tau$ or $\delta\omega_c$. (We varied $\delta\tau$ from 0 to τ_0 and changed $\delta\omega_c$ up to one order of magnitude.)

At a fixed value of the amplitude A the value of $\langle\tau_{jump}\rangle$ is approximately proportional to τ_0/Δ^2. Fig. 11.23 demonstrates this dependence.

The best fit for the numerical points in Fig. 11.23 is given by

$$\ln\langle\tau_{jump}\rangle = p + q\ln(\tau_0/\Delta^2). \tag{11.36}$$

For $A = 1.2 \times 10^5$ we have $p = 17.9$, $q = 0.993$. For the sixfold value $A = 7.2 \times 10^5$, we obtained the same value of q, and $p = 19.743$. If we estimate the amplitude of the random CT vibrations near the Rabi frequency as 1 pm, then $\Delta = 1.8$. Putting $\tau_0 = \tau_R$, we obtain $\omega_c\langle\tau_{jump}\rangle = 2.3$ s.

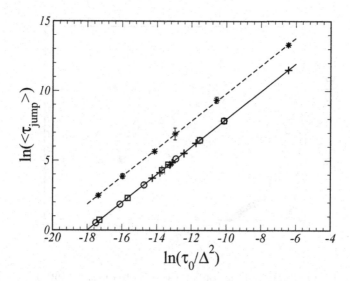

Figure 11.23: Dependence of the average time interval between two consecutive quantum jumps on τ_0/Δ^2. The full line corresponds to the value $A = 1.2 \times 10^{-5}$, the squares represent $\delta\omega_c = 4.2 \times 10^{-8}$, the crosses $\delta\omega_c = 4.2 \times 10^{-7}$, the circles $\delta\omega_c = 4.2 \times 10^{-6}$. The dashed line corresponds to the value $A = 7.2 \times 10^{-5}$. Data have been obtained by varying parameters Δ and τ_0 in the ranges: $10 < \Delta < 300$ and $0.001 < \tau_0 < 1$.

Next we computed the correlation function for the CT frequency shift

$$C(\tau_a) = \frac{\langle (\delta\omega_c(\tau) - \langle\delta\omega_c\rangle)(\delta\omega_c(\tau + \tau_a) - \langle\delta\omega_c\rangle)\rangle}{\langle(\delta\omega_c(\tau))^2\rangle - \langle\delta\omega_c\rangle^2}, \tag{11.37}$$

where $\langle\delta\omega_c\rangle = \langle\delta\omega_c(\tau)\rangle = 0$, and $\langle...\rangle$ indicates an average over time.

In Fig. 11.24, we show the correlation function $C(\tau_a)$ for three different values of parameters, as indicated in the legend. As one can see, the general behavior is well described by the exponential function (indicated by dashed lines in Fig. 11.24) $\exp(-\tau/\tau_c)$. The relation between the correlation time τ_c and $\langle\tau_{jump}\rangle$ was found to be $\langle\tau_{jump}\rangle \simeq 2.5\tau_c$.

In conclusion, we note that the quantum jumps in a single quantum

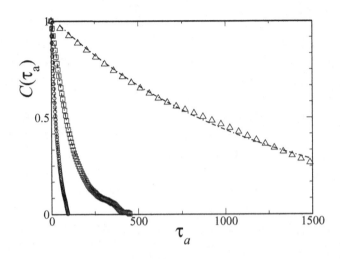

Figure 11.24: Correlation function of the CT frequency shift $C(\tau_a)$ for different parameters: circles $\Delta = 100, \tau_0 = 10^{-2}$, squares $\Delta = 50, \tau_0 = 10^{-2}$, triangles $\Delta = 50, \tau_0 = 10^{-1}$. In all cases $\delta\tau = \tau_0/4$. The dashed curves show the exponential approximation of the correlation function $\exp(-\tau/\tau_c)$ with $\tau_c = 23.91, 95.81, 1179.15$, respectively.

system are widely studied in quantum optics (see, for example, the review of Plenio and Knight [41]).

11.4   Reduction of the frequency shift due to the CT-spin entanglement

Magnetic noise on the spin causes the deviation of the spin from the direction of the effective field $\vec{B}_{eff}$. Quantum collapse "pushes" the spin back to the previous direction relative to $\vec{B}_{eff}$ or, sometimes, to the opposite direction. The latter case we call the quantum jump. During the short time interval between the two consecutive collapses the spin is entangled with the CT. It means that during this time there are two CT trajectories with the frequency shift $\delta\omega_c = \pm\delta\omega_0$, but these two trajectories are hidden inside a single "average trajectory" due to the quantum uncertainty of the CT position. The magnitude of the frequency shift for this "average trajectory" must be smaller than $\delta\omega_0$. In this section we consider this effect using a simplified model of the mysterious quantum collapse.

Below, as well as in the previous section, we use the dimensionless quantities: CT coordinate x_c and momentum $\hat{p}_c$ in "quantum units", frequency shift $\delta\omega_c$ in units of ω_c, and dimensionless time $\tau = \omega_c t$. We start from the Hamiltonian (11.24). The wave function Ψ of the CT-spin system can be written in the same form as in Eq. (8.7):

$$\Psi = \begin{pmatrix} \psi_1(x_c, \tau) \\ \psi_2(x_c, \tau) \end{pmatrix}. \qquad (11.38)$$

Note, that if the functions $\psi_1(x_c, \tau)$ and $\psi_2(x_c, \tau)$ are identical up to a constant complex factor, then the wave function Ψ can be represented by a product of the CT and the spin wave functions. In this case, the average spin $\langle \vec{S} \rangle$ has the magnitude $1/2$. In the general case, the spin is entangled with the CT, and the average spin magnitude is smaller than $1/2$.

Below we describe our computer simulations of the CT-spin dynamics with the Hamiltonian (11.24). Again, we consider the function $\Delta(\tau)$ in Eq. (11.24) as a random telegraph signal with two values $\pm\Delta$. The time interval between two consecutive "kicks" of the function $\Delta(\tau)$ was taken randomly from the interval $(3\tau_R/4, 5\tau_R/4)$. The initial wave function is chosen as a product of the CT and spin wave functions. The initial state of the CT is a coherent state (4.13) with the value of α:

$$\alpha = \frac{1}{\sqrt{2}} \left[\langle x_c(0) \rangle + i\langle p_c(0) \rangle \right]. \qquad (11.39)$$

The initial direction of the spin is taken opposite to the direction of the effective field, which is given, in units of ω_c/γ by:

$$\vec{B}_{eff} = (\epsilon, 0, -2\eta x_c). \qquad (11.40)$$

To save computational time we have used the values of parameters (10.29), where $\langle z_c(0) \rangle$ must be changed to $\langle x_c(0) \rangle$. These values of parameters correspond to the partial adiabatic reversals with the frequency shift (10.35).

In our numerical simulations the functions $\psi_1(x_c, \tau)$ and $\psi_2(x_c, \tau)$ in (11.38) have been expanded over 400 eigenfunctions of the unperturbed oscillator Hamiltonian. During the time interval between two consecutive "kicks" of the noise function $\Delta(\tau)$ we have a time–independent Hamiltonian. Thus, we find the evolution of the wave function by diagonalizing the 800×800 matrix and taking into consideration the initial conditions after each "kick". The output of our simulations is the time interval $\tau_{j+1} - \tau_j$ between two consecutive returns to the origin for the average value $\langle x_c \rangle$

$$\langle x(\tau_{j+1}) \rangle = 0 \quad \text{and} \quad \langle x(\tau_j) \rangle = 0. \tag{11.41}$$

The results of our simulations are shown in Fig. 11.25, which demonstrates the deviation of $\tau_{j+1} - \tau_j$ from the unperturbed half-period of the CT oscillations π, namely

$$\delta\tau_j = \tau_{j+1} - \tau_j - \pi. \tag{11.42}$$

Without magnetic noise ($\Delta = 0$) the deviation $\delta\tau_j$ does not change with time. In this case we have for any j:

$$\delta\tau_j = \delta\tau_0 = \pi\delta\omega_0 \simeq 0.025. \tag{11.43}$$

If $\Delta \neq 0$, then the value of $\delta\tau_j$ decreases with the increase of j (see Fig. 11.25).

Note, that we consider a relatively small time interval before the CT trajectory splits into two trajectories, i.e. before the formation of the Schrödinger cat state. We assume that the characteristic collapse time τ_{col} includes, at least, a few half-periods of the CT vibrations. Next, we assume in our model that the quantum collapse is an instantaneous event and that between the collapses, the CT-spin system evolves according to the Schrödinger equation. In this case, we may introduce the average frequency shift $\langle \delta\omega_c \rangle$:

$$\langle \delta\omega_c \rangle = -\frac{\langle \delta\tau_j \rangle}{\pi}. \tag{11.44}$$

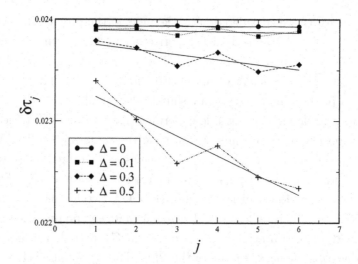

Figure 11.25: Deviation of $\tau_{j+1} - \tau_j$ from the unperturbed half period of the CT oscillations π, as a function of the number of half periods j, for different values of Δ, as indicated in the legend. Solid lines are the standard linear fits. On the vertical axis $\delta\tau_j = \tau_{j+1} - \tau_j - \pi$.

The deviation of the average magnitude of the frequency shift $\langle|\delta\omega_c|\rangle$ from the value $\delta\omega_0$ depends on the value of τ_{col}.

Note that the deviation $\langle|\delta\omega_c|\rangle$ from $\delta\omega_0$ may be interpreted as an effective reduction of the spin δS caused by the spin-CT entanglement. If we rewrite the estimation (10.35) for the partial adiabatic reversals in the form:

$$\delta\omega_c = \frac{2S\eta^2}{(2\eta^2 A^2 + \epsilon^2)^{1/2}},\tag{11.45}$$

then the reduction of the spin is given by the formula

$$\delta S = \frac{(\langle|\delta\omega_c|\rangle - \delta\omega_0)(2\eta^2 A^2 + \epsilon^2)^{1/2}}{2\eta^2}.\tag{11.46}$$

The deviation $\langle|\delta\omega_c|\rangle$ from $\delta\omega_0$ is caused by the generation of the second CT trajectory with the opposite frequency shift. During the relatively short time interval which we considered here, the peak of the probability distribution, corresponding to the second CT trajectory, is hidden inside the peak corresponding to the first CT trajectory. The second peak, generated by the magnetic noise $\Delta(\tau)$, is small compared to the first peak. That is why its contribution to the average frequency shift is also small.

Now, we will describe a possible experiment, where two CT trajectories with the opposite frequency shift $\pm\delta\omega_0$, have the same probability. We assume that during a small time interval between the quantum collapses these two peaks form a common peak with no frequency shift at all. This experiment would be close to the interrupted OSCAR technique described in Chap. 8, Sec. 2. Namely, when the CT is in one of its end points, one will turn off the *rf* field for the duration of a quarter of the CT period. It is equivalent to the application of the $\pi/2$-pulse in the RSC. The "$\pi/2$-pulse" changes the angle between the spin and the effective field. After the "$\pi/2$-pulse" we have two CT trajectories with frequencies $\pm\delta\omega_0$, each with the same probability. Before the CT-spin wave function collapses to one of these two trajectories with the corresponding direction of the spin, the CT will oscillate with the unperturbed frequency ω_c. After the collapse the frequency shift is $+\delta\omega_0$ or $-\delta\omega_0$, with equal probabilities. If one applies a periodic sequence of "$\pi/2$-pulses" with the period τ_p, and $\tau_p > \tau_{col}$, then the average frequency shift is

$$\langle |\delta\omega_c| \rangle = \frac{0 \cdot \tau_{col} + \delta\omega_0(\tau_p - \tau_{col})}{\tau_p} = \delta\omega_0 \left(1 - \frac{\tau_{col}}{\tau_p} \right). \qquad (11.47)$$

Manipulating with τ_p one could achieve a significant decrease of $\langle |\delta\omega_c| \rangle$ in comparison to $\delta\omega_0$. By measuring $\langle |\delta\omega_c| \rangle$, one can estimate from (11.47), the characteristic time of collapse of the CT-spin wave function, τ_{col}.

Chapter 12

MRFM Applications: Measurement of an Entangled State and Quantum Computation

In this chapter we will consider the applications of the MFRM techniques to the measurement of an entangled spin state and an MRFM-based nuclear spin quantum computer (Berman *et al.* [42–44]).

12.1 MRFM measurement of an entangled spin state

The entangled states are well known as the most bizarre states in quantum physics, (see, for instance, Bell's papers [45]). We will consider an entangled spin state, for example, the spin state of two particles

$$|\chi\rangle = \frac{1}{\sqrt{2}} \left(|\uparrow\uparrow\rangle + |\downarrow\downarrow\rangle \right). \tag{12.1}$$

which is described by the spin wave function

$$\chi(s_1, s_2) = \frac{1}{\sqrt{2}} \left(\alpha_1 \alpha_2 + \beta_1 \beta_2 \right), \qquad (12.2)$$

where $s_i = S_{z,i}$, $i = 1, 2$. In the entangled state (12.1) both spins have the same direction. With probability $1/2$ it may be the positive or the negative z-direction, but in any case it will be the same direction for both spins. Even if two particles are far from each other, there is some "quantum connection" between their spins. If one measures the spin z-component for the first particle and "collapses" it, say, to the state "up", i.e. $| \uparrow \rangle$, then he or she automatically collapses the spin of the second particle to the same state "up". This quantum connection cannot be used for the instantaneous transfer of the information, as a person who performs the measurement does not know in advance the result of his or her measurement.

In this section we consider the evolution of the entangled spin state (12.1) when one of the two spins is measured using MRFM. As an example, we will consider an MRFM technique with cyclic adiabatic reversals driven by the frequency modulated rf field (see Chap. 8).

We use the Hamiltonian (8.4) for the CT-spin system:

$$\mathcal{H} = \frac{1}{2} \left(\hat{p}_c^2 + z_c^2 \right) - \dot{\phi} \hat{S}_{z,1} + \epsilon \hat{S}_{x,1} - 2\eta \hat{S}_{z,1} z_c, \qquad (12.3)$$

which includes the spin operators $\hat{S}_{x,1}$ and $\hat{S}_{z,1}$ for the first spin and does not contain the operators for the remote entangled spin. The dimensionless wave function of the whole CT-spin system, including the second spin, can be written as

$$\begin{aligned} \Psi(z_c, s_1, s_2, \tau) &= \psi_{\uparrow\uparrow}(z_c, \tau)\alpha_1\alpha_2 + \psi_{\uparrow\downarrow}(z_c, \tau)\alpha_1\beta_2 \\ &+ \psi_{\downarrow\uparrow}(z_c, \tau)\beta_1\alpha_2 + \psi_{\downarrow\downarrow}(z_c, \tau)\beta_1\beta_2. \end{aligned} \qquad (12.4)$$

Substituting (12.4) into the Schrödinger equation, we derive four coupled equations for the functions $\psi(z_c, \tau)$,

$$2i\dot{\psi}_{\uparrow\uparrow} = (\hat{p}_c^2 + z_c^2 + \dot{\phi} - 2\eta z_c)\psi_{\uparrow\uparrow} - \epsilon\psi_{\downarrow\uparrow},$$

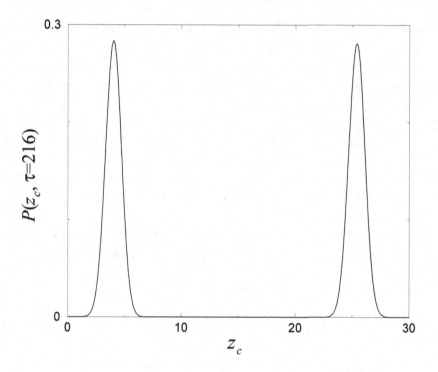

Figure 12.1: The probability distribution, $P(z_c)$, at $\tau = 216$.

$$2i\dot{\psi}_{\downarrow\uparrow} = (\hat{p}_c^2 + z_c^2 - \dot{\phi} + 2\eta z_c)\psi_{\downarrow\uparrow} - \epsilon\psi_{\uparrow\uparrow},$$

$$2i\dot{\psi}_{\uparrow\downarrow} = (\hat{p}_c^2 + z_c^2 + \dot{\phi} - 2\eta z_c)\psi_{\uparrow\downarrow} - \epsilon\psi_{\downarrow\downarrow},$$

$$2i\dot{\psi}_{\downarrow\downarrow} = (\hat{p}_c^2 + z_c^2 - \dot{\phi} + 2\eta z_c)\psi_{\downarrow\downarrow} - \epsilon\psi_{\uparrow\downarrow}. \tag{12.5}$$

This system of equations splits into two independent sets of equations. The initial state is assumed to be a product of the coherent quasiclassical state (4.13) for the CT and the entangled state (12.1) for the two spins.

Below we describe the numerical simulations for the same values of parameters as in Section (8.1) and an adiabatic increase of the *rf* field amplitude described by formula (8.17). Figure 12.1 shows the typical probability distribution of the CT position,

$$P(z_c) = |\psi_{\uparrow\uparrow}|^2 + |\psi_{\uparrow\downarrow}|^2 + |\psi_{\downarrow\uparrow}|^2 + |\psi_{\downarrow\downarrow}|^2. \tag{12.6}$$

One can see that the probability distribution, $P(z_c)$, describes a Schrödinger cat state of the CT (i.e. two CT trajectories) with two approximately equal peaks. When these two peaks are clearly separated, the total wave function can be represented as a sum of two terms corresponding to the two peaks in the probability distribution,

$$\Psi(z_c s_1, s_2, \tau) = \Psi^b(z_c, s_1, s_2, \tau) + \Psi^{sm}(z_c, s_1, s_2, \tau). \qquad (12.7)$$

Note, that the two peaks have equal amplitudes, but we prefer to use the same superscripts "b" and "sm", which we used for "big" and "small" peaks in Chaps. 8 and 10.

Our numerical analysis shows that each term, Ψ^b and Ψ^{sm}, can be approximately decomposed into a direct product of the CT and spin wave functions,

$$\Psi^b = \psi^b(z_c, \tau)\chi^b(s_1, \tau)\alpha_2, \quad \Psi^{sm} = \psi^{sm}(z_c, \tau)\chi^{sm}(s_1, \tau)\beta_2. \qquad (12.8)$$

This decomposition is possible because the complex function $\psi_{\uparrow\uparrow}(z_c, \tau)$ is proportional to $\psi_{\downarrow\uparrow}(z_c, \tau)$, and the complex function $\psi_{\uparrow\downarrow}(z_c, \tau)$ is proportional to $\psi_{\downarrow\downarrow}(z_c, \tau)$. Such proportionality can be seen in Fig. 12.2, where we plot the corresponding wave functions at the same time as in Fig. 12.1, with suitable numerical coefficients.

The spin wave function, $\chi^b(s_1, \tau)$, describes the dynamics of the first spin with its average, $\langle \chi^b | \vec{S} | \chi^b \rangle$, pointing approximately in the direction of the effective field, $\vec{B}_{eff}$, in the RSC,

$$\vec{B}_{eff} = (\epsilon, 0, -\dot{\phi}). \qquad (12.9)$$

(We neglect here the nonlinear term $2\eta z_c$ whose contribution to the effective field is small.) The spin function, $\chi^{sm}(s_1, \tau)$, describes the dynamics of the first spin with its average pointing in the direction opposite to the direction of $\vec{B}_{eff}$. As the amplitude of the CT vibrations increases, the phase difference between the oscillations of two peaks, $|\psi^b(z_c, \tau)|^2$ and $|\psi^{sm}(z_c, \tau)|^2$, approaches π. (See Fig. 12.3.) (To reach the phase difference of π, a long time of numerical simulations is required.)

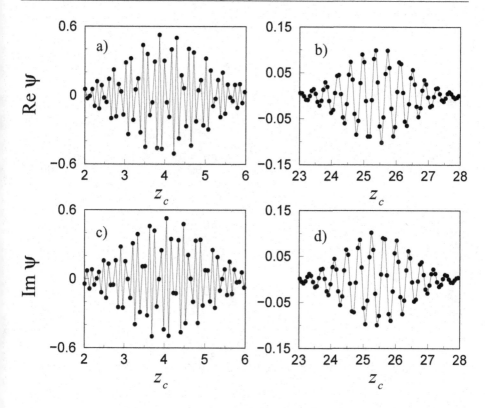

Figure 12.2: Upper boxes: the real part of wave functions; lower boxes: the imaginary part of wave functions, at $\tau = 216$. (a) The full line $Re(\psi_{\uparrow\uparrow})$, circles $Re(-5\psi_{\downarrow\uparrow})$. (b) The full line $Re(\psi_{\uparrow\downarrow})$, circles $Re(5\psi_{\downarrow\downarrow})$. (c) The full line $Im(\psi_{\uparrow\uparrow})$, circles $Im(-5\psi_{\downarrow\uparrow})$. (d) The full line $Im(\psi_{\uparrow\downarrow})$, circles $Im(5\psi_{\downarrow\downarrow})$.

In realistic experimental conditions, the Schrödinger cat state quickly collapses due to the interaction with the environment. In this case, the two peaks of the probability distribution describe two possible trajectories of the spin-CT system.

In one of these trajectories the first (measured) spin is pointed along the direction of the effective field while the second (remote) spin is pointed "up" (in the positive z-direction); the other trajectory corresponds to the opposite situation in which the orientation of both spins is reversed: the

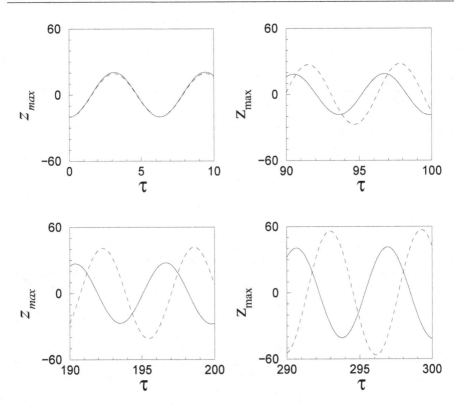

Figure 12.3: The positions, $z_{max}(\tau)$, of two peaks of the Schrödinger cat state as functions of time.

first (measured) spin is antiparallel to the effective field, and the second (remote) spin is pointed "down" (in the negative z-direction). The phase difference between the corresponding oscillations of the CT approaches π with increasing the CT vibration amplitude. This result is equivalent to a Stern-Gerlach measurement of a single spin entangled to a remote spin.

Thus, the numerical simulations reveal two possible outcomes shown schematically in Fig. 12.4: (a) The first (measured) spin points along the effective field in the RSC, and the second (remote) spin points in the positive z-direction. (b) The first (measured) spin points opposite to the direction of the effective field in the RSC, and the second (remote) spin points in the

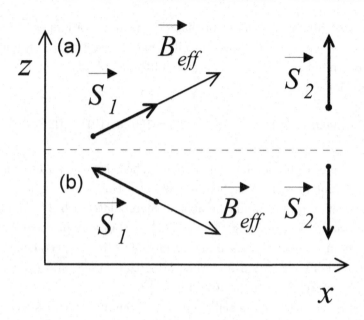

Figure 12.4: Two outcomes of the MRFM measurement of the state of two entangled spins. (a) The measured spin, $\vec{S_1}$, points along the direction of the effective field, and the remote spin, $\vec{S_2}$, points "up" (in the positive z-direction). (b) The measured spin, $\vec{S_1}$, points in the direction opposite to the effective field, and the remote spin, $\vec{S_2}$, points "down" (in the negative z-direction).

negative z-direction. Thus, the collapse of the measured spin along (or opposite) the direction of the rotating effective field leads to the collapse of the remote spin in the positive (or negative) z-direction.

12.2   MRFM based spin quantum computer

Quantum computation became one of the most popular fields in quantum physics. There are many books of various level describing the basic principles and the latest achievements in this field (see, for example, Berman *et al.* [46] and Nielsen and Chuang [47]). A quantum computer utilizes a system of quantum bits (qubits), which can be placed in a superpositional state. As an example, a system of three "classical bits" may represent $2^3 = 8$ numbers, but only one number at a time. A system of three qubits can be placed into a superpositional state

$$|\psi\rangle = \frac{1}{\sqrt{8}} \left(|000\rangle + |001\rangle + |010\rangle + |011\rangle + |100\rangle + |101\rangle + |110\rangle + |111\rangle \right),$$
(12.10)

which represents eight numbers at a time. Thus, in a quantum computer one can manipulate with all eights numbers at a time. A scalable quantum computer with N qubits would be able to manipulate with 2^N numbers at a time. It would allow solving the so called "intractable problems", which do not have an efficient "classical algorithm". Two most spectacular examples are the Shor's algorithm for prime factorization [48] and the Grover's algorithm for search over the unsorted data [49].

In order to implement a scalable quantum computer one has to create a large system of qubits with long enough relaxation and decoherence times. Next, one is expected to be able to put the qubit system into its ground state, to implement the one-qubit rotation and the Control-Not (CN) gate, and to measure the states of qubits. One-qubit rotation means an arbitrary change of the state of a qubit. The CN gate CN_{ik} is a two-qubit gate, which changes the state of the target qubit k if the control qubit i is in the state $|1\rangle$ and does not affect the state of the target qubit if the control qubit is in the state $|0\rangle$. As an example, an arbitrary state $|\psi\rangle$ of a two qubit system

$$|\psi\rangle = C_a|0_1\ 0_2\rangle + C_b|0_1\ 1_2\rangle + C_d|1_1\ 0_2\rangle + C_f|1_1\ 1_2\rangle,$$
(12.11)

under the action of the Control-Not gate CN_{12} transform into the state $|\psi'\rangle$:

$$CN_{12}|\psi\rangle = |\psi'\rangle = C_a|0_1\ 0_2\rangle + C_b|0_1\ 1_2\rangle + C_d|1_1\ 1_2\rangle + C_f|1_1\ 0_2\rangle.$$
(12.12)

Below we describe a hypothetical nuclear spin quantum computer based on MRFM. Assume that we have a chain of impurity paramagnetic atoms near the surface of a solid. Every atom has an electron spin $S = 1/2$ and a nuclear spin $I = 1/2$ and corresponding magnetic moments $\vec{\mu} = -\gamma\hbar\langle\vec{S}\rangle$ and $\vec{\mu}_n = \gamma_n\hbar\langle\vec{I}\rangle$, where $\gamma_n > 0$ is the nuclear gyromagnetic ratio. The hyperfine interaction couples electron and nuclear spins of the atom. The whole system is placed into the external permanent magnetic field $\vec{B}_{ext}$. A qubit is represented by a nuclear spin, and the two states of a qubit $|0\rangle$ and $|1\rangle$ are represented by the ground and the excited states of the nuclear spin.

First, we will discuss the measurement of the state of a nuclear spin. We assume that the sensitivity of the MRFM technique is sufficient for the measurement of the state of an electron spin but is not sufficient for the nuclear spin measurement. Let consider a simple case when the hyperfine interaction is described by the Hamiltonian

$$\mathcal{H}_{hf} = A_{hf}\,\hat{\vec{S}}\cdot\hat{\vec{I}}, \tag{12.13}$$

where A_{hf} is the positive constant of the hyperfine interaction. In an external magnetic field the frequency of the ESR will depend on the nuclear spin state. We will consider the energy levels of electron-nuclear spin system in an external magnetic field shown in Fig. 12.5.

If the ESR frequency with no hyperfine interaction is $\omega_e = \gamma B_{ext}$, then for the nuclear spin pointing "up" the ESR frequency, which we denote as ω_{e0}, is

$$\omega_{e0} = \omega_e + \omega_{hf}, \qquad \omega_{hf} = \frac{A_{hf}}{2\hbar}. \tag{12.14}$$

For the nuclear spin pointing "down" the corresponding ESR frequency ω_{e1} is

$$\omega_{e1} = \omega_e - \omega_{hf}. \tag{12.15}$$

Note, that in Fig. 12.5, we assume that the NMR frequency ω_n with no hyperfine interaction $\omega_n = \gamma_n B_{ext}$ is smaller than ω_{hf}.

A possible setup for the nuclear spin measurement in a quantum computer based on MRFM is shown in Fig. 12.6. At low temperature the electron spins are polarized and their magnetic moments point in the positive z-direction. As an example, we will consider a simple MFRM technique, where the periodic reversals of an electron spin are driven by the periodic sequence of the

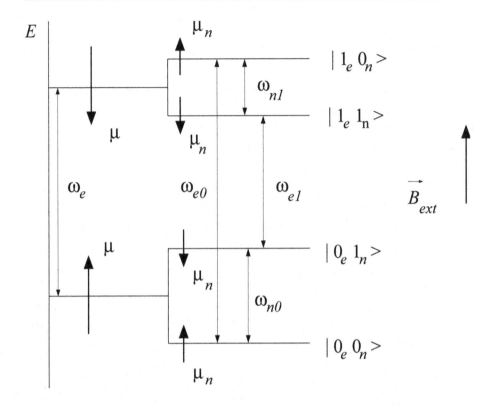

Figure 12.5: Energy levels for electron and nuclear spins 1/2 of a paramagnetic atom in a high external magnetic field. The electron and nuclear magnetic moments are indicated by μ and μ_n, ω_e and ω_n are the frequencies of the ESR and nuclear magnetic resonance (NMR) in the external magnetic field without the hyperfine interaction. The frequencies $\omega_{e0} = \omega_e + \omega_{hf}$, $\omega_{e1} = \omega_e - \omega_{hf}$, $\omega_{n0} = \omega_n + \omega_{hf}$ and $\omega_{n1} = \omega_{hf} - \omega_n$. Here ω_{hf} is the hyperfine frequency (half of the hyperfine splitting of the ESR). ω_{eq} is the ESR frequency for the nuclear state $|q\rangle$ ($q = 0, 1$). ω_{nq} is the NMR frequency for the electron state $|q\rangle$. We assume that $\omega_{hf} > \omega_n$. In the external field $B_{ext} = 10\ T$, $\omega_e \simeq 280$ GHz, and $\omega_n = \omega_n/2\pi \simeq 430$ MHz (for a proton).

resonant rf pulses (see Chap. 7). If the time interval between two consecutive π-pulses is equal to the half of the CT period T_c, then the electron spin

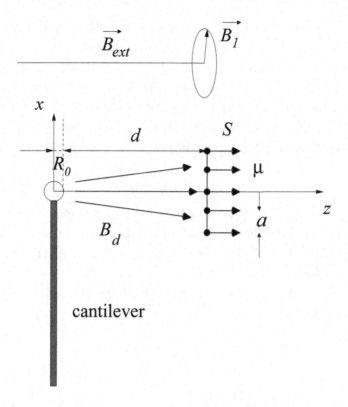

Figure 12.6: A nuclear-spin quantum computer based on MRFM. B_{ext} is the permanent external magnetic field, B_1 is the *rf* magnetic field, B_d is the nonuniform dipole magnetic field produced by a ferromagnetic particle in the sample, R is the radius of the ferromagnetic particle, d is the distance between the ferromagnetic particle and the targeted atom, and a is the distance between neighboring impurity atoms. The origin is placed at the equilibrium position of the center of the ferromagnetic particle. Arrows on the sample show the direction of the electron magnetic moments μ.

reversals will drive the resonant CT vibrations. It is clear that if the frequency of π-pulses is ω_{e0}, then the electron spin will drive the CT vibrations only if the nuclear spin points "up". If the frequency of the π-pulses is ω_{e1} the CT vibrations will be driven if the nuclear spin points "down". Thus, one can measure the state of the nuclear spin using the quantum transitions of the electron spin detected by the MRFM technique.

Suppose that the distance between the impurity atoms is $a = 5$ nm, the distance between the ferromagnetic particle and the target atom is $d = 10$ nm, and the radius of the ferromagnetic particle is $R_0 = 5$ nm (see Fig. 12.6). The dipole magnetic field at the target atom produced by the ferromagnetic particle can be estimated as

$$B_d = \frac{2}{3}\mu_0 M \left(\frac{R_0}{R_0 + d}\right)^3, \tag{12.16}$$

where M is the magnetization (magnetic moment per unit volume) of the ferromagnetic particle. Choosing $\mu_0 M = 2.2$ T, which corresponds to iron, we get $B_d = 5.4 \times 10^{-2}$ T. The corresponding shift of the ESR frequency is $\gamma B_d/2\pi = 1.5$ GHz.

Now, we estimate the z-component of the dipole field B'_{dz} at the neighboring impurity atoms:

$$B'_{dz} = \frac{1}{3}\mu_0 M \left(\frac{R_0}{r}\right)^3 \left[3\left(\frac{R_0 + d}{r}\right)^2 - 1\right], \tag{12.17}$$

$$r = \sqrt{(R_0 + d)^2 + a^2}. \tag{12.18}$$

We assume that $B_{ext} \gg B_d$, so only the z-component of the dipole field is required to estimate the ESR frequency shift. The difference between the ESR frequencies for two neighboring atoms is

$$\Delta\omega'_e = \gamma(B_d - B'_{dz}). \tag{12.19}$$

For our parameters $\Delta\omega'_e/2\pi = 500$ MHz. The electron Rabi frequency ω_R must be smaller than $\Delta\omega'_e$, to provide a selective measurement of the electron spin on the target atom.

The magnetic dipole field at the target atom oscillates due to the CT vibrations. The maximum deviation of the magnetic field can be estimated as

$$\Delta B = \left|\frac{\partial B_d}{\partial z}\right| A, \tag{12.20}$$

where A is the CT vibration amplitude. The corresponding deviation of the ESR frequency is

$$\Delta\omega_e = \gamma\Delta B. \tag{12.21}$$

As an example, for the CT amplitude $A = 0.1$ nm, the maximum deviation of the magnetic field $\Delta B = 1$ mT, and the corresponding deviation of the ESR frequency $\Delta\omega_e/2\pi = 25$ MHz. To provide spin flips, the Rabi frequency ω_R must be greater than $\Delta\omega_e$. Next, to measure a nuclear state, the Rabi frequency ω_R must be less than the hyperfine frequency ω_{hf}. Finally, the obvious condition for the π-pulses driven periodic reversals technique is $\omega_R \gg \omega_c$. Thus, the final requirement for the Rabi frequency can be written as

$$\omega_c, \Delta\omega_e \ll \omega_R \ll \Delta\omega_e', \omega_{hf}. \tag{12.22}$$

For our parameters the Rabi frequency $\omega_R/2\pi \approx 100$ MHz, roughly satisfies the inequalities (12.22).

We should estimate also the dipole field $\vec{B}_d^*$ on the atom produced by the other paramagnetic atoms. The main contribution to the dipole field is associated with the neighboring atoms. For any inner atom in the chain, two neighboring electron magnetic moments which point in the positive z-direction produce the dipole magnetic field $\vec{B}_{d1}^*$ of magnitude

$$B_{d1}^* = \frac{2\mu_0\mu_B}{4\pi a^3} \approx 1.5 \times 10^{-5} \text{ T}. \tag{12.23}$$

The maximal contribution from all other paramagnetic atoms to the dipole field $|B_{d2}^*|$ does not exceed 3×10^{-6} T. (For a chain of 1000 paramagnetic atoms with electron spins in the ground state, the value of B_{d2}^* at the center of the chain is $B_{d2}^* \approx 0.202 B_{d1}^*$.) The corresponding frequency shift of the ESR is

$$\left(\frac{\gamma}{2\pi}\right)|B_{d1}^* + B_{d2}^*| \leq 500 \text{ kHz}. \tag{12.24}$$

This frequency shift is negligible compared to the estimated electron Rabi frequency $\omega_R/2\pi \approx 100$ MHz. Thus, to measure the nuclear-spin state of the target atom, one can use a periodic sequence of "electron" π-pulses with frequency

$$\omega = \omega_{e0} + \gamma B_d. \tag{12.25}$$

Next, we will discuss the operation of the MRFM quantum computer. First, the single-spin MRFM can be used to create 100% polarization in a nuclear-spin chain. To do this, one should determine the initial state of each nuclear spin in the chain. Note, that we assume 100% polarization of electron spins. In the external magnetic field $B_{ext} = 10$ T, at the temperature $T \approx 1$ K, the probability for an electron spin to occupy the upper energy level is approximately

$$\exp(-2\mu_B B_{ext}/k_B T) \approx 1.4 \times 10^{-6}.$$

During the measurement process, one should use an even number of pulses to return the electron spin to the ground state.

To create 100% polarization of the nuclear spins or to carry out a quantum computation, one must fix the z coordinate of the ferromagnetic particle. This nonvibrating particle is not a measuring device. It is only a static source of the inhomogeneous magnetic field. One can imagine that the ferromagnetic particle could move along the spin chain (see Fig. 12.7). It can be used to target each nuclear spin which is in the excited state. According to Eqs. (12.16) and (12.17) the target nuclear spin experiences an additional magnetic field $B_d \approx 5.4 \times 10^{-2}$ T. A neighboring nuclear spin experiences an additional magnetic field $B'_{dz} \approx 3.6 \times 10^{-2}$ T. The corresponding shifts of the NMR frequencies are $(\gamma_n/2\pi)B_d \approx 2.3$ MHz and $(\gamma_n/2\pi)B'_{dz} \approx 1.5$ MHz. (Here we present estimates for a proton, $(\gamma_n/2\pi) \approx 4.3 \times 10^7$ Hz/T). The frequency difference between the target nuclear spin and its neighbor is, $\Delta\omega'_n/2\pi \approx 800$ kHz.

In order to drive a nuclear spin one may use an rf pulse, which is resonant to the nuclear spin. The frequency of this pulse is equal to the NMR frequency, which is much smaller than the ESR frequency. If we denote the rotating rf field driving the nuclear spin as $\vec{B}_{n1}$, then the nuclear Rabi frequency ω_{nR} is

$$\omega_{nR} = \gamma_n B_{n1}. \tag{12.26}$$

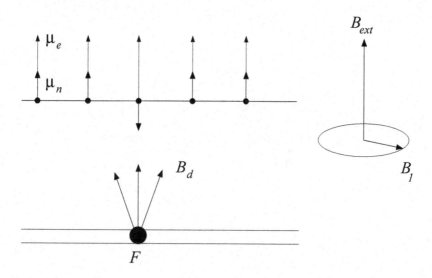

Figure 12.7: The polarization of nuclear spins: The non-vibrating ferromagnetic particle targets a nuclear spin which is initially in the excited state.

The nuclear Rabi frequency ω_{nR} must be smaller than $\Delta\omega'_n$, in order to provide a selective action of a nuclear π-pulse.

The dipole field B^*_{d1} produced by two neighboring paramagnetic atoms (for inner nuclear spins in the chain) is given by Eq. (12.23). The maximal additional contribution of all other paramagnetic atoms was estimated to be $|B_{dz}| < 3 \times 10^{-6}$ T. The corresponding shift of the NMR frequency is approximately 780 Hz, much less than the assumed value of the Rabi frequency, $\omega_{nR} \approx 100$ kHz. Thus, to drive the target nuclear spin into its ground state, one can apply a "nuclear" π-pulse with the frequency

$$\omega = \omega_{n0} + \gamma_n B_d. \qquad (12.27)$$

In this way, the whole chain of the nuclear spins can be initialized in its ground state. In the same way, using various rf pulses instead of a single π-pulse, one can provide a one-qubit rotation for any selected nuclear spin.

Now we consider the possibility of implementing conditional logic in a chain of nuclear spins. The direct interaction between nuclear spins for in-

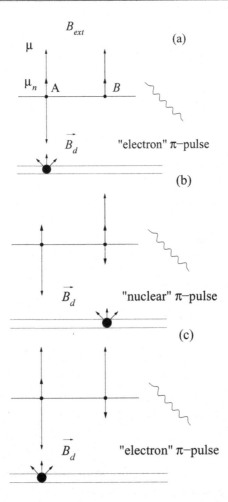

Figure 12.8: Implementation of a quantum CN gate: (a) an "electron" π-pulse drives the electron magnetic moment of the control qubit (nuclear spin A) if A is in the ground state; (b) a "nuclear" π-pulse causes a transition in the target qubit (nuclear spin B) if the control qubit A is in the ground state; (c) "electron" π-pulse drives the electron magnetic moment back into the ground state.

teratomic distance $a = 5$ nm is negligible. That is why, to provide conditional logic in a system of nuclear spins, one will use an electron dipole field. Suppose that we want to implement a two-qubit quantum CN logic gate. We will consider, first, a simpler "inverse" CN gate: the target qubit changes its state if the control qubit is in the ground state. The target qubit is a nuclear spin which can change its state during the CN operation, but not necessarily the nuclear spin closest to the ferromagnetic particle.

Assume that the target qubit is any inner nuclear spin in the chain, and the control qubit is one of its neighboring nuclear spins. (See Fig. 12.8, where A and B are the control and target qubits.) We want to implement an "inverse" CN gate in three steps.

(1) One sets the ferromagnetic particle near the control qubit, Fig. 12.8(a), and applies an "electron" π-pulse with frequency given by Eq. 12.25. The electron Rabi frequency ω_R satisfies the inequalities $\omega_R < \Delta\omega'_e, \omega_{hf}$. So, the electron magnetic moment of the left paramagnetic atom in Fig. 12.8 changes its direction only if the control qubit is in the ground state $|0\rangle$. (2) The ferromagnetic particle moves to the target qubit, see Fig. 12.8(b). If the control qubit is in the excited state, then the electron magnetic moment of the left atom did not change its direction during the first step. In this case, the NMR frequency for the target qubit is

$$\gamma_n \left(B_d + B_{d1}^* + B_{d2}^* \right). \tag{12.28}$$

The second term in the sum is important for us $(\gamma/2\pi)B_{d1}^* \approx 650$ Hz. The value $(\gamma/2\pi)B_{d2}^*$ depends on the position of the nuclear spin in the chain. We estimated that the range of variation for this term is approximately between 70 Hz and 130 Hz. The exact value of this term can be calculated or measured experimentally for each nuclear spin in the chain. If the control qubit is in the ground state (as in Fig. 12.8), then the electron magnetic moment of the left atom changed its direction during the first step. In this case, the NMR frequency of the target qubit is

$$\gamma_n \left(B_d + B_{d2}^* \right), \tag{12.29}$$

because the dipole field produced by neighboring paramagnetic atoms cancels out: $B_{d1}^* = 0$.

Next, one applies a "nuclear" π-pulse with frequency (12.29). The difference between the frequencies in Eq. (12.28) and (12.29) is approximately 650 Hz. Thus, the nuclear Rabi frequency $\omega_{nR}/2\pi$ must be less than 650 Hz. The corresponding duration of the "nuclear" π-pulse is $\tau > 770$ μs. Under the action of a "nuclear" π-pulse the target qubit changes its state if the control qubit is in its ground state.

(3) To complete the CN gate, the ferromagnetic particle moves back to the control qubit, see Fig. 12.8(c). Then one should again apply the "electron" π-pulse with frequency (12.25). This pulse drives the electron magnetic moment back to its ground state. A similar procedure can be applied if the target qubit is at either end of the chain. In order to implement a "direct" CN-gate instead of the "inverse" CN-gate one must apply a "nuclear" π-pulse of frequency (12.28) instead of (12.29).

Next, we will consider a concrete physical system, which can be used as an MRFM quantum computer. It may be, for example, a silicon substrate with tellurium impurities. A tellurium atom in silicon is a "deep"donor with a small electron cloud and with an extremely large hyperfine interaction. Application of tellurium impurities in silicon could combine the advantages of MRFM with the well-developed techniques of silicon technology. We consider ^{125}Te nuclei with a spin $I = 1/2$, whose natural abundance is only 7%. Suppose that the regular chain of ^{125}Te impurities is placed near the surface of the ^{28}Si substrate. ^{28}Si nuclei are nonmagnetic $(I = 0)$. The atoms with ^{29}Si magnetic nuclei whose natural abundance is 4.7% are supposed to be removed.

When the host atom in silicon is replaced by the tellurium donor, two extra electrons become available. The properties of tellurium donors in silicon can be found, for example, in the article of Grimmeiss et al. [50]. It was found that most of the implanted tellurium atoms occupy substitutional sites. The ground state of the tellurium donors, as well as those of other atoms with two extra electrons, are referred to as "deep impurity levels", in contrast to "shallow" impurities like phosphorus with one extra electron whose ground state energies are of the order 50 meV. Because of the two extra electrons, tellurium donors form singly ionized A centers, Te^+, and neutral B centers, Te^0. The temperature-independent ground state energies were found to be 410.8 meV for A centers, and 198.8 meV for B centers.

Unlike the typical case, considered above, the gyromagnetic ratio for ^{125}Te is negative , like the electron gyromagnetic ratio. The "spin Hamiltonian" of the A-centers, which are supposed to be used in a quantum computer, can be written as:

$$\mathcal{H} = \gamma \hbar \vec{B} \cdot \hat{\vec{S}} + \gamma_n \hbar \vec{B} \hat{\vec{I}} - A \hat{\vec{S}} \hat{\vec{I}}, \tag{12.30}$$

where $\gamma_n/2\pi = 13.45$ MHz/T and $A/2\pi\hbar = 3.5$ GHz.

Below we describe the initialization to the ground state, one-qubit rotation and the CN gate for this particular system. Let assume that the external magnetic field $B_{ext} = 10$ T, and the temperature is 1 K. At these conditions the electron spins are polarized, but 44% of nuclear spins (qubits) are in their excited states. To detect these nuclear spins one moves a cantilever to every tellurium atom one by one.

Applying the periodic sequence of the rf π-pulses with frequency (12.25) ($\omega/2\pi = 283.25$ GHz) one drives the periodic reversals of the electron spin and the CT vibrations only if the nuclear spin is in the ground state. If the applied π-pulses do not drive the CT vibrations then it indicates that the corresponding nuclear spin is not in the ground state. In this case, one applies a "nuclear π-pulse" of frequency (12.27) $\omega/2\pi \approx 1.885226$ GHz, which drives this nuclear spin into its ground state. Using various rf pulses of the same frequency one can implement one-qubit rotations.

To implement a CN-gate one moves the non-vibrating ferromagnetic particle to a tellurium ion containing a control nuclear spin (a control qubit). Then, one applies a π-pulse with frequency (12.25). This pulse drives the electron spin into the excited state if the control nuclear spin is in its ground state. Next, one moves the non-vibrating ferromagnetic particle to the neighboring tellurium ion containing the target nuclear spin (a target qubit) and applies a "nuclear" π-pulse of frequency (12.28) $\omega/2\pi \approx 1.885230$ GHz. This pulse changes the state of the target nuclear spin if the dipole contribution from neighbor electron spins does not cancel out. This happens only if the control nuclear spin was in the excited state. Finally, one moves the non-vibrating ferromagnetic particle back to the ion containing the control nuclear spin and applies a π-pulse with the frequency (12.25) to return the electron spin to its ground state (if it had been in the excited state). Thus, three rf pulses together implement a quantum CN-gate: the target qubit changes its state if the control qubit is in the excited state. Note, that the final measurement of the nuclear states can be implemented using MRFM in the same way as the measurement of the initial nuclear states.

Chapter 13

MRFM Techniques and Spin Diffusion

Application of any spin device crucially depends on the spin relaxation time. Spin diffusion, associated with "flip-flops" between the neighboring spins caused by the dipole-dipole interaction, is one of the most important factors in a spin relaxation process.

The idea of spin diffusion originated from Bloembergen who explained nuclear spin-lattice relaxation in insulating crystals [51]. He demonstrated that the transport of magnetization from fast relaxing spins (FRS) to slow relaxing spins (SRS) can be described as a diffusion process. Due to the spin diffusion, a small amount of FRS (e.g. located near the impurities) can greatly accelerate the spin-lattice relaxation in the whole spin system.

Budakian et al. [52] have shown that the high gradients of the magnetic field used in MRFM apparatus can be used for effective suppression of the spin diffusion. In this chapter we will describe the theory of this effect and its possible application for a spin quantum computer (Berman et al. [53, 54]).

13.1 Spin diffusion in the presence of a nonuniform magnetic field

According to the general theory of the spin diffusion in the ESR (Vugmeister [55]) the relaxation process depends on the relation between two parameters: the spin-lattice relaxation time for FRS T_{FL} and the cross relaxation time T_{FS}, which is the characteristic time of the energy transfer from FRS to SRS. We will consider the case $T_{FS} \ll T_{FL}$.

If $T_{FS} \ll T_{FL}$ then FRS and SRS quickly come to the state of thermal equilibrium, and the bottleneck of the relaxation process is the energy transfer from FRS to the lattice. In this case the overall relaxation rate T_1^{-1} is given by the parameter T_0^{-1}:

$$T_1^{-1} \approx T_0^{-1} = (n_F/n_S)T_{FL}^{-1}, \qquad (13.1)$$

where n_F and n_S are the concentrations of FRS and SRS. The spin-lattice relaxation of SRS with the characteristic time T_{SL} is ignored in this expression.

We will consider a quasiclassical electron "spin" with magnetic moment $\vec{\mu}_p$. The magnitude of the magnetic moment, which is equal to the Bohr's magneton $\mu_p = \mu_B$, conserves in the process of the spin relaxation. The motion of a magnetic moment $\vec{\mu}_p$ satisfies to the quasiclassical equation of motion with the relaxation term $\vec{R}_p$:

$$\dot{\vec{\mu}}_p = -\gamma[\vec{\mu}_p \times \vec{B}_p] + \vec{R}_p,$$

$$\vec{R}_p = \frac{\xi_p}{\mu_B}[\vec{\mu}_p \times \dot{\vec{\mu}}_p]. \qquad (13.2)$$

Here $\vec{B}_p$ is the magnetic field on spin "p", which includes the uniform external magnetic field $\vec{B}_{ext}$, the nonuniform dipole magnetic field produced, for example, by a non-vibrating ferromagnetic particle $\vec{B}_{dp}$ and the dipole field produced by other spins of a sample $\vec{B}_{dp}^*$; ξ_p is the relaxation parameter. We assume that $\vec{B}_{ext}$ points in the positive z-direction.

The dipole field $\vec{B}_{dp}^{*}$ is given by

$$\vec{B}_{dp}^{*} = \frac{\mu_0}{4\pi} \sum_{k \neq p} \frac{3(\vec{\mu}_k \cdot \vec{n}_{kp})\vec{n}_{kp} - \vec{\mu}_k}{r_{kp}^3}, \tag{13.3}$$

where $\vec{n}_{kp}$ is the unit vector, which points from the spin k to the spin p, r_{kp} is the distance between the two spins. Below we use two conditions:

$$B_{dp}^{*}, \; \left| \vec{B}_{dp} - \left\langle \vec{B}_d \right\rangle \right| \ll B_{ext} + \left\langle B_{dz} \right\rangle$$

where $\langle ... \rangle$ means the average over the spin system.

Note, that according to the Maxwell's equation, $div\,\vec{B}_d = 0$, there must be a nonuniform transversal component of the magnetic field $\vec{B}_d$. However, because of $\left| \vec{B}_{dp} - \left\langle \vec{B}_d \right\rangle \right| \ll B_{ext} + \left\langle B_{dz} \right\rangle$ the transversal component of $\vec{B}_d$ in the first approximation does not influence the Larmor frequency.

The average Larmor frequency ω_0 in the spin system can be found approximately as

$$\omega_0 = \gamma\langle B_0\rangle, \; \langle B_0\rangle = B_{ext} + \langle B_{dz}\rangle, \tag{13.4}$$

We transfer to the system of coordinates, which rotates with the frequency ω_0. Ignoring the fast oscillating terms and assuming $\xi_p \ll 1$ we obtain the following equations of motion:

$$\frac{d\vec{\mu}_p}{d\tau'} = [\vec{\Omega}_p \times \vec{\mu}_p], \tag{13.5}$$

$$\vec{\Omega}_p = \left(-\frac{\chi}{2} \sum_{k \neq p} A_{kp}\mu_{kx} + \xi_p'\mu_{py}, \; -\frac{\chi}{2} \sum_{k \neq p} A_{kp}\mu_{ky} - \xi_p'\mu_{px}, \; \delta_p + \chi \sum_{k \neq p} A_{kp}\mu_{kz} \right).$$

Here we use the following dimensionless notations:

$$\dot{\vec{\mu}}_p = d\vec{\mu}_p/d\tau', \; \tau' = \xi_0\omega_0 t, \; \xi_p' = \xi_p/\xi_0, \tag{13.6}$$

$$\delta_p = \frac{B_{dpz} - \langle B_{dz}\rangle}{\xi_0\langle B_0\rangle}, \; \chi = \frac{\mu}{4\pi} \frac{\mu_B}{\xi_0 a^3 \langle B_0\rangle}, \; A_{kp} = \frac{3\cos^2\theta_{kp} - 1}{(r_{kp}/a)^3},$$

where the magnetic moment $\vec{\mu}_p$ is taken in units of μ_B, a is the average distance between the spins, θ_{kp} is the polar angle of the unit vector $\vec{n}_{kp}$, $\xi_0\omega_0 = 1\text{s}^{-1}$.

If we put origin inside the spin system at the point where $\delta_p = 0$, then the value of δ_p can be approximated as

$$\delta_p = \frac{aG}{\xi_0 \langle B_0 \rangle} \left(\frac{z_p}{a} \right), \qquad G = \frac{\partial B_d}{\partial z} \bigg|_{z=0}. \tag{13.7}$$

The parameter χ is the characteristic constant of the dipole-dipole interaction in our system, while the ratio $\tilde{G} = aG/(\xi_0 \langle B_0 \rangle)$ describes the characteristic Larmor frequency difference caused by the magnetic field gradient.

The approximations leading to (13.5) from (13.2) maintain, as it should be, the conservation of the magnetic moment:

$$\mu_{px}^2 + \mu_{py}^2 + \mu_{pz}^2 = 1. \tag{13.8}$$

In the absence of relaxation ($\xi_p' = 0$) the z-component of the total magnetic moment also conserves: $\sum_p \dot{\mu}_{pz} = 0$.

In experiments [52], the values of parameters are the following:

$$B_0 = 0.106 \text{ T}, \qquad \omega_0/2\pi = 2.96 \text{ GHz}, \qquad a \approx 8 \text{ nm},$$

$$G \approx 10 \text{ T/m} \div 36 \text{ kT/m}, \qquad T_1 \approx 6.25 \text{ s} \div 25 \text{ s}.$$

The value of χ is $\chi = 3.2 \times 10^5$. When G changes from 10 T/m to 36 kT/m the dimensionless parameter $\tilde{G}$ rises from 0.044χ to 156χ for $a = 8$ nm. Thus, the Larmor frequency difference for the neighboring spins becomes greater than the dipole–dipole interaction constant.

The computational scheme for Eq. (13.5) must satisfy to the conditions of conservation μ_p and (in the absence of relaxation) $\sum_p \mu_{pz}$. Below we describe the suggested computational algorithm. Equations (13.5) are approximated by the difference equations

$$\frac{\vec{\mu}_p^{j+1} - \vec{\mu}_p^j}{\Delta \tau'} = \left[\vec{\Omega}_p \left(\vec{\mu}_1^{(av)}, \vec{\mu}_2^{(av)}, ... \vec{\mu}_N^{(av)} \right) \times \frac{\vec{\mu}_p^{j+1} + \vec{\mu}_p^j}{2} \right], \tag{13.9}$$

where $\vec{\mu}_k^{(av)} = (\vec{\mu}_k^{j+1} + \vec{\mu}_k^j)/2$, $k = 1, 2, ...N$, N- number of magnetic moments in a system, $\Delta \tau' = \tau_{j+1}' - \tau_j'$, j- counts the time instant. Equation (13.9) can be written as

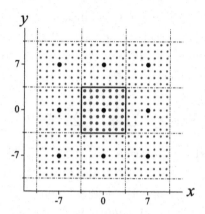

Figure 13.1: A pattern of the $x - y$ plane containing FRS for $n = 7$. The values of x and y are given in units of a. $\bullet -$ FRS.

$$\left(\vec{\mu}_p^{j+1}\right)_i = \sum_l f_{il}(\vec{\Omega}_p) \cdot \left(\vec{\mu}_p^j\right)_l. \tag{13.10}$$

Here $i, l = x, y, z$, $f_{il}(\vec{\Omega}_p)$- are nonlinear functions of $\vec{\Omega}_p$. As parameters $\vec{\Omega}_p$ depend on $\vec{\mu}_k^{j+1}$ the solution of equations of motion was found by iterations.

In our simulations, we consider the system of identical cells (see Fig. 13.1). Every cells contains $N = n^3$ spins located near the points of the cubic lattice:

$$\vec{r}_p = a(l_{p1} + l_{p2} + l_{p3}) + \delta \vec{r}_p.$$

Here n is an odd number, l_p- are integers, $-(n-1)/2 \leq l_{p1,2,3} \leq (n-1)/2$, δx_p, δy_p, δz_p- are random numbers, which do not exceed $0.05a$. At $\tau' = 0$, the z-components of $\vec{\mu}_p$ take random values between -0.8 and -1 with the random direction of the transversal components. In every cell, the central spin is the only FRS. In our simulations, we take the same value $\xi'_p = \tilde{\beta}$ for all FRS and the same value $\xi'_p = \tilde{\alpha}$ for all SRS.

The main (central) cell is surrounded by 440 identical auxiliary cells in order to eliminate the boundary effects, which influence the relaxation process.

We computed the dynamics of the spins in the central cell taking into consideration their interaction with the spins of all the cells and assuming that the corresponding spins in all cells have the same direction. The outcome of our computation is the function

$$M_z(\tau') = \sum_p \mu_{pz}(\tau') \sum_p |\mu_{pz}(0)|. \tag{13.11}$$

We have performed computations for $n = 5, 7, 11$. The corresponding ratio $n_s/n_f = N - 1 = 124, 342, 1330$. First, we have studied the relaxation process in the absence of magnetic field gradient ($\tilde{G} = 0$). We define the effective dimensionless relaxation time τ_1 using the relation

$$\tau_1 = \frac{1}{2} \int\limits_0^\infty [1 - M_z(\tau')] \, d\tau', \tag{13.12}$$

which is derived from the exponential decay $M_z(\tau') = 1 - 2\exp(-\tau'/\tau_1)$. As $\xi_0\omega_0 = 1$ s^{-1}, the numerical value of τ_1 is equal to the value T_1 in seconds. For the spin system with no FRS the relaxation time $T_1 = T_{SL} \approx 2.2/\tilde{\alpha}$ (in seconds). In experiment [52] $T_{SL} = 25$ s, and $\tilde{\alpha} = 0.088$.

We have simulated the relaxation process for various values $\chi > 10^4$, but all our figures below correspond to the experimental value $\chi = 3.2 \times 10^5$.

Our computations show that the overall relaxation time τ_1 can be described by the relation

$$\tau_1 = \frac{2.2}{\tilde{\alpha} + \tilde{\beta}/N}. \tag{13.13}$$

Putting $N \approx n_s/n_F$ and transferring to the dimensional relaxation rate, we obtain from (13.13)

$$T_1^{-1} = T_{SL}^{-1} + T_0^{-1}. \tag{13.14}$$

The first term in this expression describes the spin-lattice relaxation rate for SRS, and the second term $T_0^{-1} = (n_F/n_s)T_{FL}^{-1}$, first derived in [55], describes the effect of the spin diffusion. From the experimental data [52] $T_1^{-1} =$

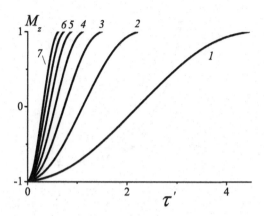

Figure 13.2: Spin relaxation $M_z(\tau')$ for the various values of the ratio $\tilde{\beta}/N$ and $\tilde{\alpha} = 1$. Curves 1-7: $\tilde{\beta}/N = 0$, 1, 2, 3, 4, 5, 6.

0.16 s^{-1}, $T_{SL}^{-1} = 0.04$ s^{-1} we can estimate the value of T_0^{-1}: $T_0^{-1} = 0.12$ s^{-1}. The corresponding value $\tilde{\beta}/N \approx 0.26$.

Figure 13.2 demonstrates the dependence $M_z(\tau')$ for various values of the ratio $\tilde{\beta}/N$.

Figure 13.3 shows the dependence of the relaxation time τ_1 on the ratio $\tilde{\beta}/N$ for $\tilde{\alpha} = 0$ and $\tilde{\alpha} = 1$. One can see the excellent agreement between the numerical data and formula (13.13).

Next, we transfer to the main objective of this section: the analysis of the influence of the magnetic field gradient on the relaxation process. Our simulations show that the suppression of the relaxation rate depends on the single parameter $K = \tilde{G}/\chi$, which is the ratio of the characteristic Larmor frequency difference to the dipole-dipole interaction constant. The two other results of our simulations are the following:

1. The significant increase of the relaxation time τ_1 appears in the region $4.4 \leq K \leq 44$, which corresponds to the values 1 kT/m $\leq G \leq 10$ kT/m, in a good agreement with experiments [52].

2. For $K > 44$, the ratio $R(K) = \tau_1(K)/\tau_1(K = 0)$ is approaching the

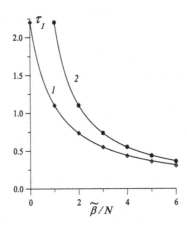

Figure 13.3: The overall relaxation time τ_1 as a function of ratio $\tilde{\beta}/N$. 1- $\tilde{\alpha} = 1$; 2- $\tilde{\alpha} = 0$. Dots are received from the numerical calculations, solid line corresponds to formula (13.13).

value $1 + \tilde{\beta}/(\tilde{\alpha}N)$. It means that the overall relaxation time $T_1(K)$ is approaching the expected value T_{SL}.

As an example, Fig. 13.4 demonstrates the dependence $M_z(\tau')$ for 5 values of K, and Fig. 13.5 shows the function $R(K) = \tau_1(K)/\tau_1(K = 0)$.

Next, we have studied the dynamics of the spatial distribution $\mu_{pz}(\vec{r}_p, t)$. We have found that in the absence of the magnetic field gradient ($K = 0$), the relaxation process spreads randomly in all directions from FRS to SRS. In the presence of the magnetic field gradient, the spin diffusion process becomes anisotropic. Figure 13.6 demonstrates the relaxation of a slice magnetic moment M_{zi} for all the slices, $-(n-1)/2 \leq i \leq (n-1)/2$, in the central cell. One can see that the relaxation process first develops in the slice containing FRS ($i = 0$), then it spreads to the slices $i < 0$ below the central slice, then it spreads to the upper slices $i > 0$. This phenomenon can be explained as following.

The spin frequency in the rotating frame is given by Eq. (13.5). We have

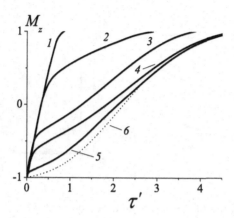

Figure 13.4: The relaxation $M_z(\tau')$ for various values of K: 1–5 — $K = 0, 5, 10,$ 20, 50. Curve 6 is the relaxation in the spin system with no FRS and $K = 0$. The values of other parameters: $\tilde{\alpha} = 1$, $n = 7$, $\tilde{\beta} = 1372$ ($\tilde{\beta}/N = 4$).

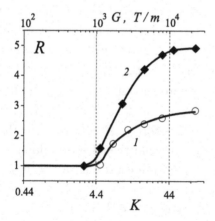

Figure 13.5: Dependence R on K at $\tilde{\alpha} = 1$, $n = 7$. Curves 1, 2 — $\tilde{\beta} = 1372, 686$ ($\tilde{\beta}/N = 4,\ 2$).

$$\Omega_p \approx K\chi(z_p/a) + \Omega_{pd}, \qquad \Omega_{pd} = \chi \sum_{k \neq p} A_{kp}\mu_{kz} , \qquad (13.15)$$

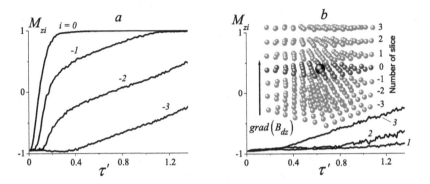

Figure 13.6: Relaxation of the slice magnetic moment $M_{zi}(\tau')$ in the presence of the magnetic field gradient: (a) for the central slice, which contain FRS ($i = 0$), andfor the slices below the central slice; (b) for the slices above the central slice. The values of parameters: $n = 7$, $\tilde{\alpha} = 1$, $\tilde{\beta}/N = 4$, $K = 10$.

where the first term, $\propto K$, is associated with the nonuniform dipole magnetic field produced by the ferromagnetic particle, and the second term, Ω_{pd}, is associated with the dipole field produced by other spins of the system.

Within the slice the value of Ω_{pd} changes slightly between 5.6χ and 6.4χ. Thus, the value Ω_p is, approximately, $\Omega_0 = 6\chi$ in the slice $i = 0$, $\Omega_1 = (6 + K)\chi$ in the slice $i = +1$, $\Omega_{-1} = (6 - K)\chi$ in the slice $i = -1$, and so on. For $\Omega_1 - \Omega_0 = \Omega_0 - \Omega_{-1} \geq \Omega_0$ (or $K > 6$) the relaxation due to the spin diffusion develops first in the slice containing FRS ($i = 0$). When the magnetic moment M_{z0} in the central slice, $i = 0$, changes its direction, $M_{z0} \approx -1 \rightarrow M_{z0} \approx +1$, the dipole contribution on the neighboring slices $i = \pm 1$ in our disturbed cubic lattice changes as

$$\Delta\Omega_i \approx 0.35\chi[1 + M_{z0}(\tau)], \ i = \pm 1. \tag{13.16}$$

The dipole term Ω_{dp} within the central slice itself changes as

$$\Delta\Omega_0 \approx -9\chi[1 + M_{z0}(\tau)] . \tag{13.17}$$

Thus, after the spin relaxation in the slice $i = 0$, the frequency difference between the slices decreases for the slices $i = 0$ and $i = -1$, but increases for $i = 0$ and $i = 1$. As a result, the spin relaxation process due to spin diffusion starts in the slice $i = -1$, then in the slice $i = -2$, and so on.

The characteristic value $K = K_0$, which is necessary to suppress the spin diffusion between the slices, can be estimated as following. We assume that at $K = K_0$ the frequency difference between the slices $i = 0$ and $i = -1$ after the spin relaxation in the slice $i = 0$ remains greater than $\Omega_0 = 6\chi$. Taking into consideration that the initial frequency difference between these slices is $K_0\chi$, the change of the frequency is -18χ for $i = 0$ and 0.7χ for $i = -1$, we obtain $K_0 \approx 25$, which is in a good agreement with our numerical simulations.

13.2 Suppression of the spin diffusion in a spin quantum computer

In this section we discuss how the effect of the suppression of the spin diffusion could be used for a spin quantum computer. We will consider two schemes for a spin quantum computer. In the first one a boundary spin chain in a two-dimensional (2D) spin array is used as a one-dimensional (1D) spin quantum computer. In the second one an isolated spin chain is used as a 1D quantum computer. While the chain of spin qubits is supposed to be free from FRS, an FRS may appear at some distance from the chain. We will use the same equations of motion (13.5) and the same notation as in the previous section.

We will start from the first scheme for the quantum computer described above. We will discuss spin diffusion and relaxation in a 2D array of spins and its influence on the boundary chain of spins. We consider the following initial conditions for the relaxation process: at $\tau' = 0$ the z-components of $\vec{\mu}_p$ are given random values between -0.8 and -1 with a random direction for the transverse components. The spin coordinates in the $x - y$ plane are

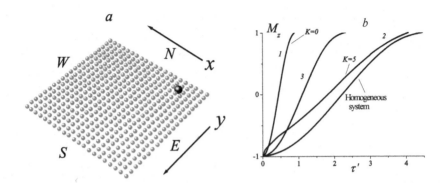

Figure 13.7: (a) a 2D spin system with FRS. FRS is shown as a big sphere, SRS are represented by small spheres. Spin quantum computer is the boundary spin chain, which is parallel to the x-axis. $G_z > 0$. (b) curves 1 and 3 show the dependence $M_z(\tau')$ for $K = 0$: 1- for $\tilde{\alpha} = 1$, $\tilde{\beta}/N = 4$; 3 – for $\tilde{\alpha} = 0$, $\tilde{\beta}/N = 2$. Curve 2 corresponds to $K = 5$, $\tilde{\alpha} = 1$, $\tilde{\beta}/N = 4$. The "homogeneous system" curve corresponds to $K = 0$, $\tilde{\alpha} = \beta = 1$. For all curves $N = 441$ ($n = 21$).

given by

$$\vec{r}_p = a(l_{p1}, l_{p2}, 0) + \delta\vec{r}_p,$$

$$\delta\vec{r}_p = (\delta x_p, \delta y_p, 0), \qquad (13.18)$$

where $-(n-1)/2 \leq l_{p1,2} \leq (n-1)/2$, n is an odd number (in order to be able to put an FRS in the center of the 2D plane), the total number of spins is $N = n^2$. δx_p, δy_p- are random numbers, which do not exceed $0.05a$.

We consider a single FRS with the relaxation parameter $\xi_p' = \tilde{\beta}$. All SRS have the relaxation parameter $\xi_p' = \tilde{\alpha} \ll \tilde{\beta}$. We assume that the dipole magnetic field B_{dpz} produced by the ferromagnetic particle increases with the increase of x, and we approximate the value of δ_p as

$$\delta_p = \frac{aG_z}{\xi_0 B_0}\left(\frac{x_p}{a}\right), \quad G_z = \left.\frac{\partial B_{dpz}}{\partial x}\right|_{x=0}. \qquad (13.19)$$

Our simulations show that similar to the three-dimensional (3D) case the suppression of the relaxation rate depends on the ratio $K = \tilde{G}/\chi$, where $\tilde{G} = aG_z/(\xi_0\langle B_0\rangle)$. Figure 13.7(a) demonstrates one of the studied 2D spin systems with FRS. Our simulations show that at $K = 0$ the relaxation process in the whole 2D-spin system does not depend on the location of the FRS. The relaxation time τ_1 defined in (13.12) is described by the same expression (13.13) as in the case of a 3D system. In Fig. 13.7(b), curves 1 and 3 demonstrate the dependence $M_z(\tau')$ for $K = 0$. Note, that for $K = 0$, the spin relaxation for all boundary spin chains (N, S, W, E in Fig. 13.7b) is approximately the same as relaxation of the whole spin system (see Fig. 13.8).

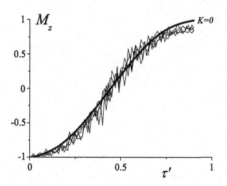

Figure 13.8: Dependence $M_z(\tau')$ for $K = 0$. Smooth solid line corresponds to the whole spin system, broken lines correspond to boundary chains N, S, W and E in Fig. 13.7(a). $\tilde{\alpha} = 1$, $\tilde{\beta}/N = 4$, $N = 441$.

We believe that this result is valid for cases in which the number of spins N does not exceed the threshold value $N_c = n_c^2$, which depends on the effective constant χ of the dipole–dipole interaction.

A rough estimate for n_c can be obtained in the following way. If the FRS is located at the center of a 2D spin system, the dipole–dipole interaction between the FRS "q" and the most remote spin "k" must satisfy the condition

$$\Omega_{kz}^{(q)} = \chi A_{kq} \gg 2\pi/\tau_1 \quad \text{or} \quad 8\chi/n_c^3 \gg \pi/\tau_1, \tag{13.20}$$

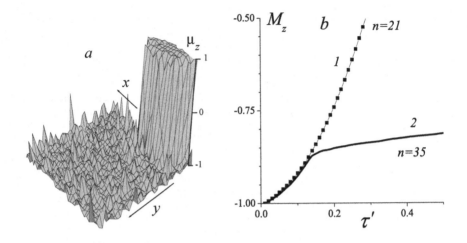

Figure 13.9: (a) Formation of a magnetic domain in a large spin system. $\tau' = 0.5$, $M_z = -0.865$. Parameters: $\chi = 1.6 \times 10^5$ ($a = 48$ nm), $\tilde{\beta}/N = 4$, $N = 1225$ ($n = 35$). (b) Curve 2 demonstrates the dependence $M_z(\tau')$, and curve 1 is identical to curve 1 in Fig. 13.7.

in order for Eq. (13.13) to be valid. If this condition is not fulfilled, the spin diffusion becomes inhomogeneous over the spin system. The SRS which are located near the FRS form a magnetic domain (see Fig. 13.9). At the boundary of this domain, the dipole field $\vec{B}_{dp}^*$, produced by the spins, has a sufficient gradient to suppress the spin diffusion outside the domain.

The effect of reducing the spin relaxation rate due to the magnetic field gradient can be described as following. Consider, for example, three spin chains parallel to the y-axis. In Fig. 13.10, they correspond to the rows -1, 0, and $+1$. Let B_{dz} increases in the positive x-direction ($G_z = \partial B_{dz}/\partial x|_{x=0} > 0$). The dimensionless Larmor frequency of spin p can be written as

$$\Omega_p \approx K\chi \left(\frac{x_p}{a}\right) + \Omega_{pd}, \qquad \Omega_{pd} = \chi \sum_{k \neq p} A_{kp}\mu_{kz}. \qquad (13.21)$$

The initial value of Ω_{pd} (when all $\mu_{pz} \approx -1$) in the 2D square spin lattice

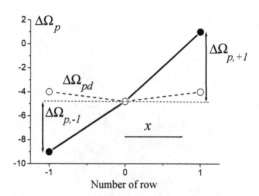

Figure 13.10: Change of the dimensionless Larmor frequencies in the neighboring rows -1, 0, 1. FRS is in the row "0". Circles show the change of frequencies due to the dipole–dipole interaction between the spins. Black circles show the frequency change due to the magnetic field gradient at $K = 5$. $\Delta\Omega_{p,-1} < \Delta\Omega_{p,+1}$.

is $\Omega_{pd} \approx 8\chi$. Suppose that row "0" contains a single FRS. Initially the relaxation process occurs in row "0". The value of μ_{pz} in row "0" sharply changes to $\mu_{pz} \approx +1$. The corresponding changes of Ω_{pd} in rows 0, 1, and -1 are approximately the same:

$$\Delta\Omega_{pd,0} \approx -4.8\chi, \quad \Delta\Omega_{pd,\pm 1} \approx -4\chi. \tag{13.22}$$

Spin diffusion is suppressed if the Larmor frequency difference due to magnetic field gradient becomes greater than $\Omega_{pd,0}$. This occurs roughly at $K > 4$. Our numerical simulations confirm this conclusion (see Fig. 13.7(b), curve 2). At $K = 10$ the dependence $M_z(\tau')$ is approximately the same as for the system with no FRS. (See Fig. 13.7(a), "homogeneous" curve.)

In a nonuniform magnetic field, spin relaxation is highly anisotropic. The relaxation process spreads first in the direction of the smaller magnetic field (because, as our simple example in Fig. 13.10 shows, $\Delta\Omega_{p,-1} < \Delta\Omega_{p,+1}$), as it occurred in 3D system. This phenomenon can be used in a quantum computer. By changing the direction of the magnetic field gradient one can

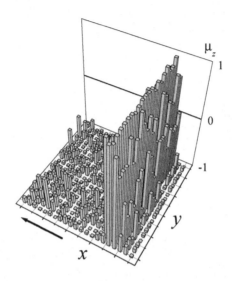

Figure 13.11: Distribution of μ_z at $\tau' = 0.35$ for the relaxation process shown in Fig. 13.7(b), curve 2.

increase the relaxation time in any of the boundary chains (N, S, W, E in Fig. 13.7). As an example, Fig. 13.11 demonstrates the distribution of μ_z at $\tau = 0.35$ for the relaxation process shown in Fig. 13.7(b), curve 2.

Next, we will consider a quantum computer implemented on an isolated spin chain. First, we consider a spin chain with $N = 41$ spins interacting with the FRS. The external magnetic field is perpendicular to the chain. We used two values $\chi = 1.3 \times 10^6$ and $\chi = 1.6 \times 10^5$, which correspond to the distances $a = a_0 = 24$ nm and $a = 2a_0 = 48$ nm. We put $\tilde{\alpha} = 0$, and $\tilde{\beta}/(N + 1) = 5$. For these values of parameters in 3D and 2D spin systems we calculated the relaxation time using Eq. (13.13), $\tau_1 \approx 2.2(N + 1)/\tilde{\beta} = 0.44$. If the FRS is at the distance $a_d \approx a$ from the center of the chain (see Fig. 13.12(a)), then the relaxation rate is similar to that in 3D and 2D systems. Figure 13.12(b) demonstrates the relaxation $M_z(\tau)$ in the spin chain for the situation shown in Fig. 13.12(a). Figure 13.12(c) demonstrates the dynamics of the FRS. (The value of the magnetic moment of the FRS μ_{qz} was

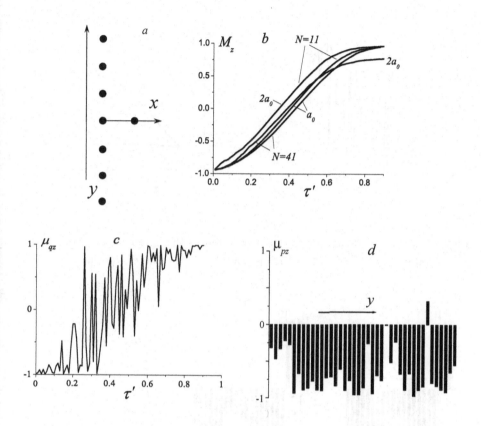

Figure 13.12: (a) FRS is placed at a distance $a_d \approx a$ from the center of the chain; (b) Relaxation $M_z(\tau')$ in the spin chain for the number of spins in the chain $N = 11$ and $N = 41$. The distance between the spins is equal to a_0 and $2a_0$ ($a_0 = 24$ nm); (c) relaxation of the FRS $\mu_{qz}(\tau')$; (d) distribution of μ_{pz} in the spin chain for $\tau' = 0.2$. $N = 41$, $a = a_0 = 24$ nm.

taken with the time interval $\Delta\tau' = 0.01$.) Figure 13.12(d) demonstrates the random distribution of the magnetic moments μ_{pz} in the spin chain at $\tau' = 0.2$ ($M_z = -0.68$).

Note, that the dynamics of a single spin in the chain is similar to the dynamics of the FRS. As an example, Fig. 13.13(a) shows relaxation of the

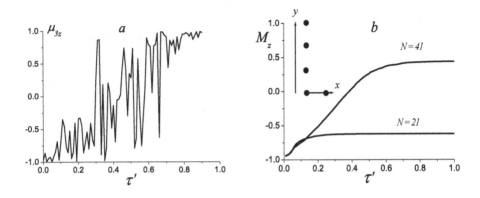

Figure 13.13: (a) Relaxation of the third spin in the spin chain. $N = 41$; (b) relaxation process $M_z(\tau')$ when FRS is placed near the end of the chain.

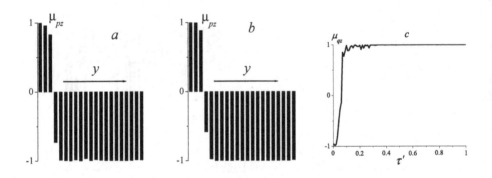

Figure 13.14: (a) and (b) Magnetic moments distribution at $\tau' = 0.3, 1$ for the spin chain with $N = 21$. (c) Magnetic moment of FRS μ_{qz} as a function of τ'.

third spin in the chain, $\mu_{3z}(\tau')$. If the FRS is placed near the end of the spin chain, the relaxation process in the spin chain slows down, and stops. (See Fig. 13.13(b).) This "freezing" of the relaxation process is accompanied by the appearance of the stationary domain walls in the chain. (See Fig. 13.14 for $N = 21$ and Fig. 13.15 for $N = 41$.) For $N = 21$ only those spins which

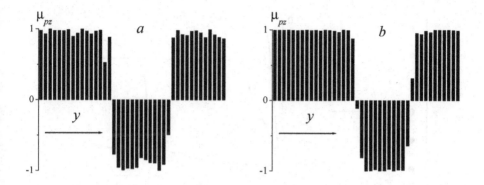

Figure 13.15: (a) and (b) Magnetic moments distribution at $\tau' = 0.6, 1$ for the spin chain with $N = 41$.

are close to the FRS take part in the relaxation process. The relaxation process freezes with two domain walls for $N = 41$.

Next, we consider the case in which the FRS and the chain spins which are close to FRS experience a higher spin dipole field than the other spins in the chain (Fig. 13.16(a), positions 1 and 2). Figure 13.16(b) demonstrates the relaxation process for the three positions of the FRS shown in Fig. 13.16(a). One can see that the high inhomogeneity of the dipole field near the FRS suppresses the relaxation rate.

Figure 13.17 shows distributions of the magnetic moments corresponding to position 1 of FRS in Fig. 13.16(a) at two different instants of time. The distribution of the magnetic moments for the FRS in position 2 (see Fig. 13.16(a)) is shown in Fig. 13.18. Note that in both Fig. 13.17 and 13.18 the distribution of the magnetic moments appears to be random rather than ordered.

For position 3 in Fig. 13.16(a) the spin dipole field on the FRS is much smaller than the spin dipole field on the spins of the chain. The relaxation rate is suppressed initially but quickly increases at $\tau' \approx 6$. (See curve 3 in Fig. 13.16(b).) The characteristic feature of this case is the generation of a moving domain wall. (See Fig. 13.19.) We have found that the direction of

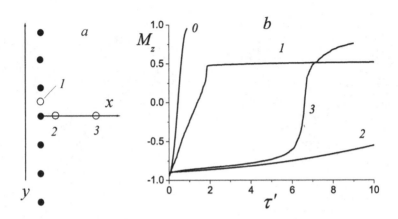

Figure 13.16: (a) FRS is very close to the spin chain (positions 1 and 2). In position 2, the distance from the chain $a_d = a/2$; in position 3, $a_d = 2a$. (b) Relaxation of the magnetic moment of the chain $M_z(\tau')$ for three positions of the FRS shown in (a). Curve "0" is taken from Fig. 13.12b for comparison. $N = 21$.

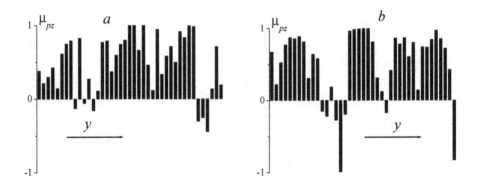

Figure 13.17: (a) and (b) Magnetic moments distributions corresponding to FRS position 1 in Fig. 13.16(a) at $\tau' = 3, 7$.

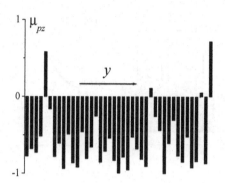

Figure 13.18: Magnetic moments distributions corresponding to position 2 of FRS in Fig. 13.16(a) at $\tau' = 10$.

the motion of the domain wall depends on the position of FRS. If we place the FRS to the left of the chain, then the domain wall will move in the negative y-direction in Fig. 13.16(a). Thus, the process of spin relaxation in the spin chain appears to generate the random spin distribution, the quasi-stationary domain walls, and the moving domain wall.

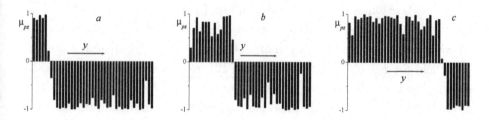

Figure 13.19: (a, b and c) "Moving domain wall" corresponding to position 3 of FRS in Fig. 13.16(a) at $\tau' = 6, 6.5, 7$.

Finally, we consider the suppression of the spin diffusion and relaxation using a nonuniform magnetic field. As an example we will discuss the situation corresponding to an FRS in location 2 in Fig. 13.16(a). Fig. 13.20(a)

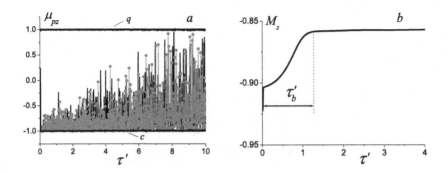

Figure 13.20: (a) Relaxation process for FRS in position "2" in Fig. 13.16(a). Horizontal lines "c" and "q" correspond to the central spin and FRS. Chaotic lines correspond to other spins of the chain. $N = 41$, $K = 0$. (b) $M_z(\tau')$ for FRS in position "2" in Fig. 13.16(a). $N = 41$, $K = 50$. Magnetic field increases in the positive x-direction.

clarifies the spin dynamics in the uniform magnetic field for this situation. The central spin "c", which is close to the FRS, quickly approaches the stable inverted state $\mu_{cz} = -1$, while FRS approaches the stable ground state $\mu_{qz} = +1$ (horizontal lines "c" and "q" in Fig. 13.20(a)). For other spins, their magnetic moments μ_{pz} change chaotically. Figure 13.20(b) demonstrates the relaxation of $M_z(\tau')$ in the spin chain in the presence of the nonuniform magnetic field for $K = 50$. In this case the FRS magnetic moment m_{qz} quickly approaches the value $\mu_{qz} = +1$, the central spin "c" also approaches the ground state $\mu_{cz} = +1$, while the other spins in the chain remain frozen. The change of $M_z(\tau')$ in Fig. 13.20(b) is caused by the relaxation of the central spin during the time interval τ'_b.

For a spin quantum computer, one should suppress the relaxation of the central spin. Relaxation of the FRS changes the dipole field on the neighboring spins of the chain to the value -2χ. In order to suppress relaxation effectively the frequency difference between the spins in the chain must increase after FRS relaxation. Consequently, the magnetic field should increase

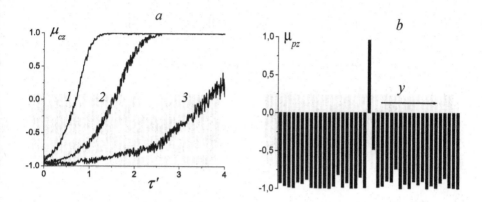

Figure 13.21: (a) Relaxation of the central spin for $K = 50$ (1), $K = 70$ (2), $K = 100$ (3). (b) Distribution of magnetic moments at $\tau' = 10$ for $K = 100$. (FRS is placed in the position "2" in Fig. 13.16(a), $N = 41$).

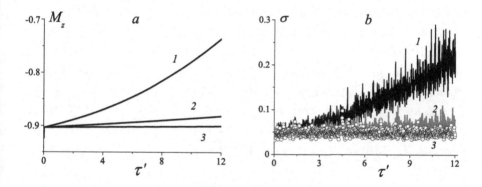

Figure 13.22: (a) Relaxation in the chain for $a_d = a$, $K = 5$(curve 1), 10 (curve 2), 25 (curve 3). (b) mean square deviation $\sigma(\tau') = \left[\frac{1}{N}\sum_p (\mu_{pz}(\tau') - M_z(\tau'))^2\right]^{1/2}$ as a function of τ' for $K = 5$, 10 and 25.

in the positive x-direction in Fig. 13.16(a). Figure 13.21(a) shows the relaxation of the central spin for three values of K, and Fig. 13.21(b) demonstrates

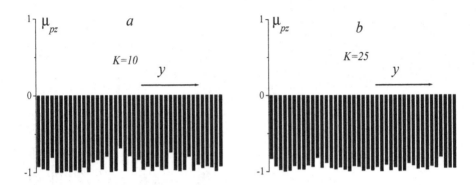

Figure 13.23: Distribution of magnetic moments at $\tau' = 12$ for $K = 10$ (a) and $K = 25$ (b).

the magnetic moment distribution for $K = 100$. If the distance a_d between FRS and the central spin increases, then the FRS can be "isolated" at smaller values of K. In the previous figures we used $a_d = a/2$. Figure 13.22 demonstrates the results of our simulations for $a_d = a$. One can see that for $K \geq 10$ the relaxation process is suppressed for every individual spin. Figure 13.23 demonstrates the distribution of the magnetic moments for $K = 10$ and $K = 25$ at $\tau' = 12$. Our semi-empirical estimate for the value of K, which is sufficient for the suppression of relaxation for the location of FRS on the x-axis in Fig. 13.16(a), can be written as $K(a_d/a)^4 > 10$.

Chapter 14

Conclusion

The purpose of our book was to explain the basic principles and some theoretical approaches used in MRFM. We did not want to consider or even to list all proposed MRFM techniques and applications. Instead, we tried to present a logic sequence of examples, which will help readers with a variety of backgrounds understand the physical principles underlying MRFM techniques and applications.

We mentioned already that our book is based largely on the research work of the authors. Anyone interested in other directions of MRFM research can read the reviews of Sidles *et al.* [56], Nestle *et al.* [57], Suter [58], the dissertation of Kriewall [59], and also recent research articles. We will now mention few articles which reflect some of the recent developments in MRFM.

Brun and Goan [60] have derived the stochastic quantum state diffusion equation, which describes the process of the wave function collapse for the MRFM technique with the adiabatic spin reversals driven by a frequency modulated *rf* field. For the same MRFM technique Gassmann *et al.* [61] have found the evolution of the CT and spin reduced density matrices for any temperature. Kempf and Marohn [62] have presented a method for obtaining the $2D$ spin density map in MRFM. Garner *et al.* [63] have developed a modified MRFM technique which relies on the magnetic force gradient rather than on the magnetic force. Budakian *et al.* [64], using the interrupted

OSCAR technique demonstrated the creation of spin order in small ensembles of electron spins. With the same technique Mamin *et al.* [65] reported an MRFM sensitivity of about 2000 net nuclear spins ^{19}F in CaF_2 and ^1H in biomolecules.

Finally, as Rugar *et al.* noted in their paper [8] the MRFM sensitivity has been increased by a factor of 10^7 compared to the first experiments [5] reported in 1992. The further improvement by a factor of 10^3 would allow a direct single-nuclear-spin detection. It would open the way for MRI with an atomic scale resolution. On this optimistic note we conclude our book.

14.1 Abbreviations

1D, 2D, 3D — one-, two-, three dimensional

AFM — atomic force microscopy

CN — Control-Not

CT — cantilever tip

ESR — electron spin resonance

FDMR — fluorescence detected magnetic resonance

FMR — ferromagnetic resonance

FRS — fast relaxing spin

LSC — laboratory system of coordinates

MFM — magnetic force microscopy

MRFM — magnetic resonance force microscopy

MRI — magnetic resonance imaging

NMR — nuclear magnetic resonance

rf — radio frequency

rms — root mean square

RSC — rotating system of coordinate

SRS — slow relaxing spins

STM — scanned tunneling microscopy

14.2   Prefixes

G (giga) $= 10^9$

M (mega) $= 10^6$

k (kilo) $= 10^3$

m (milli) $= 10^{-3}$

μ (micro) $= 10^{-6}$

n (nano) $= 10^{-9}$

p (pico) $= 10^{-12}$

f (femto) $= 10^{-15}$

a (atto) $= 10^{-18}$

z (zepto) $= 10^{-21}$

14.3   Notations

$$\alpha \quad = \begin{pmatrix} 1 \\ 0 \end{pmatrix}$$

$|\alpha\rangle$ — coherent state of the oscillator

$$\tilde{\alpha} \quad = \xi'_p \quad \text{for SRS}$$

$$\beta \quad = \begin{pmatrix} 0 \\ 1 \end{pmatrix}$$

$$\tilde{\beta} \quad = \xi'_p \quad \text{for FRS}$$

γ — electron gyromagnetic ratio (magnitude)

γ_n — nuclear gyromagnetic ratio (magnitude)

$\delta\omega_c$ — CT frequency shift

$\delta\omega_0$ — maximum value of $|\delta\omega_c|$

δk_c — shift of the CT spring constant

$$\delta_p \quad = \frac{B_{dpz} - \langle B_{dz} \rangle}{\xi_0 \langle B_0 \rangle}$$

δT_c — shift of the CT period

δz_c — uncertainty in the CT position

$$\Delta \quad = \omega_b / 2\pi$$

$$\epsilon \quad = \omega_R / \omega_c$$

η — dimensionless parameter of the spin — CT interaction

$\vec{\mu}$ — magnetic moment of an electron spin

μ_0 — permeability of the free space

$\vec{\mu}_n$ — magnetic moment of a nuclear spin

μ_B — Bohr's magneton

$\mu_\pm$ = $\mu_x \pm i\mu_y$

ω — frequency of the rf field

ω_0 — average Larmor frequency

ω_b — bandwidth of the measuring device

ω_c — "unperturbed" CT frequency

ω_e = γB_{ext} — ESR frequency in the external magnetic field

ω_{eq} — ESR frequency for the state $|q\rangle$ of the nuclear spin, $q = 0, 1$

ω_{eff} = γB_{eff} — precession frequency in the RSC

ω_{hf} = $\dfrac{A_{hf}}{2\hbar}$

ω_j — eigenfrequency of the cantilever

ω_L — Larmor frequency in MRFM

ω_n = $\gamma_n B_{ext}$ — NMR frequency in the external magnetic field

ω_{nq} — NMR frequency for the state $|q\rangle$ of the electron spin, $q = 0, 1$

ω_R = γB_1 — Rabi frequency

Ψ — wave function of the CT-spin system

$\hat{\chi}$ — spin density matrix

χ — spin wave function

χ = $\dfrac{\mu_0}{4\pi} \dfrac{\mu_B}{\xi_0 a^3 \langle B_0 \rangle}$

θ_{eff} — polar angle of the effective field

ξ_0 $= \dfrac{1s^{-1}}{\omega_0}$, where ω_0 is in s^{-1}

ξ_p — relaxation parameter for the spin "p"

ξ_p' $= \dfrac{\xi_p}{\xi_0}$

τ $= \omega_c t$ — dimensionless time

$\vec{\tau}$ — torque

τ' $= \xi_0 \omega_0\, t$

τ_1 $= \dfrac{1}{2} \displaystyle\int_0^\infty [1 - M_z(\tau')]d\tau'$

τ_{col} $= \omega_c\, t_{col}$

τ_{jump} $= \omega_c\, t_{jump}$

τ_m — spin relaxation time in the RSC

a — average distance between spins

$\hat{a}$ — annihilation operator

$\hat{a}^\dagger$ — creation operator

a_d — distance between the FRS and the center of the spin chain

A — CT amplitude

A_{hf} — hyperfine constant

A_R^T — amplitude of the thermal CT vibrations

$B_\pm$ $= B_x \pm iB_y$

$\vec{B}_0$ $= \vec{B}_{ext} + \vec{B}_d^{(0)}$

$\vec{B}_1$ — rf— field

$\vec{B}_{1n}$ — rf— field in NMR

$\vec{B}_d$ — dipole field on the spin produced by the CT

$\vec{B}_d^{(0)}$ — dipole field on the spin produced by the CT in the equilibrium position

$\vec{B}_d^*$ — dipole field on the spin produced by other spins on the sample

d — distance between the bottom of the ferromagnetic particle and the spin

d^* — distance between the origin and the spin

$\hat{E}$ — unit matrix

$\vec{F}$ — force on the CT

G $= \dfrac{\partial B_d}{\partial z}$

$\tilde{G}$ $= \dfrac{aG}{\xi_0 \langle B_0 \rangle}$

G_a $= \dfrac{\partial B_a}{\partial x}, a = x, y, z$

$\mathcal{H}$ — Hamiltonian

$\vec{I}$ — nuclear spin

k_B — Boltzmann constant

k_c — CT spring constant

K $= \dfrac{\tilde{G}}{\chi}$

l_c — cantilever length

$\vec{m}$ — magnetic moment of the ferromagnetic particle on the CT

$$m^* = \frac{k_c}{\omega_c^2} \text{ — CT effective mass}$$

m_c — cantilever mass

M_0 — magnetization of the ferromagnetic particle (magnitude)

$$M_z = M_z(\tau') = \frac{\sum_p \mu_{pz}(\tau')}{\sum_p |\mu_{pz}(0)|}$$

$n = N^{1/3}$ — for 3D systems,

$\quad = N^{1/2}$ — for 2D systems

$\vec{n}$ — unit vector

n_F — concentration of FRS

n_s — concentration of SRS

$n_\pm = n_x \pm in_y$

N — number of spins

P — probability distribution for the CT position

Q — quality factor

$$R = R(K) = \frac{\tau_1(K)}{\tau_1(K = 0)}$$

R_0 — radius of the ferromagnetic particle

$\hat{R}_j$ — rotational operator

$\vec{S}$ — electron spin

$\hat{S}_\pm = \hat{S}_x \pm i\hat{S}_y$

$s = S_z$

t_c — cantilever thickness

t_{col} — characteristic collapse time

t_{jump} — characteristic time between two quantum jumps

T — temperature

$T_0^{-1} = \dfrac{n_f}{n_s} T_{FL}^{-1}$

T_{FL} — spin-lattice relaxation time for FRS

T_{SL} — spin-lattice relaxation time for SRS

T_1 — spin relaxation time

T_c — CT period

T_d — decoherence time

T_i — period of interruption of the rf field

$T_r = \dfrac{Q}{\omega_c}$ — time constant (relaxation time) of the CT vibrations

T_R — Rabi period

u_n — eigenfunctions of the oscillator Hamiltonian

u_α — eigenfunctions of the annihilation operator $\hat{a}$.

U_m — magnetic energy

w_c — cantilever width

x_c, z_c — CT — coordinates

Y — Young's modulus

z_0 — CT — equilibrium position

Bibliography

[1] L. Ciobanu, D. A. Seeber and C. H. Pennington, J. Magn. Reson. **158**, 178 (2002).

[2] A. Blank, C. R. Dunnam, P. P. Borbat and J. H. Freed, J. Magn. Reson. **165**, 116 (2003).

[3] J. A. Sidles, Appl. Phys. Lett. **58**, 2854 (1991).

[4] J. A. Sidles, Phys. Rev. Lett. **68**, 1124 (1992).

[5] D. Rugar, C. S. Yannoni and J. A. Sidles, Nature **360**, 563 (1992).

[6] D. Rugar, O. Züger, S. Hoen, C. S. Yannoni, H. M. Vieth and D. Kendrick, Science **264**, 1560 (1994).

[7] Z. Zhang, P. C. Hammel and P. E. Wigen, Appl. Phys. Lett. **68**, 2005 (1996).

[8] D. Rugar, R. Budakian, H. J. Mamin, B. W. Chui, Nature **430**, 329 (2004).

[9] J. Kohler, Phys. Rep. **310**, 261 (1999).

[10] F. Jelezko, T. Gaebel, I. Popa, A. Gruber and J. Wrachtrup, Phys. Rev. Lett. **92**, 076401 (2004).

[11] Y. Manassen, I. Mukhopadhyay and N. R. Rao, Phys. Rev. B **61**, 16223 (2000).

[12] J. M. Elzerman, R. Hanson, L. H. Willems van Beveren, B. Witkamp, L. M. K. Vandersypen and L. P. Kouwenhoven, Nature **430**, 431 (2004).

[13] M. Xiao, I. Martin, E. Yablonovitch and H. W. Jiang, Nature **430**, 435 (2004).

[14] L. D. Landau and E. M. Lifshitz, *Theory of Elasticity*, Pergamon, New York (1986).

[15] G. P. Berman, G. D. Doolen, P. C. Hammel and V. I. Tsifrinovich, Phys. Rev. A **65**, 032311 (2002).

[16] G. P. Berman, F. Borgonovi, G. V. Lopez and V. I. Tsifrinovich, Phys. Rev. A **68**, 012102 (2003).

[17] T. D. Stowe, K. Yasumura, T. W. Kenny, D. Botkin, K. Wago and D. Rugar, Appl. Phys. Lett. **71**, 288 (1997).

[18] Sh. Kogan, *Electronic Noise and Fluctuations in Solids*, Cambridge University Press, Cambridge (1997).

[19] A. O. Caldeira and A. T. Leggett, Physica A **121**, 587 (1983).

[20] W. G. Unruh and W. H. Zurek, Phys. Rev. D **40**, 1071 (1989).

[21] B. L. Hu, J. P. Paz and Y. Zhang, Phys. Rev. D **45**, 2843 (1992).

[22] G. P. Berman, and V. I. Tsifrinovich, Phys. Rev. B **61**, 3524 (2000).

[23] L. D. Landau and E. M. Lifshitz, *Statistical Physics* , Pergamon, London (1958).

[24] G. P. Berman, F. Borgonovi, G. Chapline, S. A. Gurvitz, P. C. Hammel, D. V. Pelekhov, A. Suter and V. I. Tsifrinovich, J. Phys. A **36**, 4417 (2003).

[25] G. P. Berman, F. Borgonovi, Hsi-Sheng Goan, S. A. Gurvitz and V. I. Tsifrinovich, Phys. Rev. B **67**, 094425 (2003).

[26] B. C. Stipe, H. J. Mamin, C. S. Yannoni, T. D. Stowe, T. W. Kenny, and D. Rugar, Phys. Rev. Lett. **87**, 277602 (2001).

[27] H. J. Mamin, R. Budakian, B. W. Chui, and D. Rugar, Phys. Rev. Lett. **91** 207604 (2003).

[28] G. P. Berman, F. Borgonovi, V. N. Gorshov and V. I. Tsifrinovich, IEEE transactions on Nanotechnology **4**, 14 (2005).

[29] G. P. Berman, D. I. Kamenev and V. I. Tsifrinovich, Phys. Rev. A **66**, 023405 (2002).

[30] G. P. Berman, F. Borgonovi and V. I. Tsifrinovich, Quantum Inform. Comput. **4**, 102 (2004).

[31] G. P. Berman, F. Borgonovi and V. I. Tsifrinovich, Phys. Rev. B **72** 224406 (2005).

[32] N. N. Bogolubov and Y. A. Mitropolsky *Asymptotic methods in the theory of non-linear oscillations*, Delhi, Hindustan Pub. Corp. (1961).

[33] G. P. Berman, F. Borgonovi and V. I. Tsifrinovich, Phys. Lett. A **331** 187 (2004).

[34] G. P. Berman, F. Borgonovi and V. I. Tsifrinovich, Phys. Lett. A **337** 161 (2005).

[35] G. P. Berman, V. N. Gorshkov, D. Rugar and V. I. Tsifrinovich, Phys. Rev. B **68** 094402 (2003).

[36] G. P. Berman, V. N. Gorshkov and V. I. Tsifrinovich, Phys. Lett. A **318** 584 (2003).

[37] G. P. Berman, V. N. Gorshkov and V. I. Tsifrinovich, J. Appl. Phys. **96**, 5081 (2004).

[38] G. P. Berman, V. N. Gorshkov and V. I. Tsifrinovich, Phys. Rev. B **69**, 212408 (2004).

[39] D. Mozyrsky, I. Martin, D. Pelekhov and P. C. Hammel, Appl. Phys. Lett. **82**, 1278 (2003).

[40] B. W. Chui, Y. Hishinuma, R. Budakian, H. J. Mamin, T. W. Kenny and D. Rugar, presented at the 12th International Conference on Solid-State Sensors, Actuators and Microsystems, June 9-12, 2003, Boston, Massachusetts.

[41] M. B. Plenio, P. L. Knight, Rev. Mod. Phys. **70**, 101 (1998).

[42] G. P. Berman, F. Borgonovi, G. Chapline, P. C. Hammel and V. I. Tsifrinovich, Phys. Rev. A **66**, 32106 (2002).

[43] G. P. Berman, G. D. Doolen, P. C. Hammel and V. I. Tsifrinovich, Phys. Rev. B **61**, 14694 (2000).

[44] G. P. Berman, G. D. Doolen, P. C. Hammel and V. I. Tsifrinovich, Phys. Rev. Lett. **86**, 2894 (2001).

[45] J. S. Bell, *Speakable and Unspeakable in Quantum Mechanics: Collected Papers on Quantum Philosophy*, Cambridge University Press, Cambridge (1987).

[46] G. P. Berman, G. D. Doolen, R. Mainieri and V. I. Tsifrinovich, *Introduction to Quantum Computers*, World Scientific, Singapore- New Jersey- London- Hong Kong (1998).

[47] M. A. Nielsen and I. L. Chuang *Quantum Computation and Quantum Information*, Cambridge University Press, (2000).

[48] P. Shor, Proceedings of the 35th Annual Symposium on the Foundations of Computer Science. IEEE, Computer Society Press, New York, (1994), p. 124.

[49] L. K. Grover, Phys. Rev. Lett. **79**, 325 (1997).

[50] H. G. Grimmeiss, E. Janzen, H. Ennen, O. Schirmer, J. Schneider, R. Worner, C. Holm, E. Sirtl and P. Wagner, Phys. Rev. B **24**, 4571 (1981).

[51] N. Bloembergen, Physica (Utrecht) **15**, 386 (1949).

[52] R. Budakian, H. J. Mamin and D. Rugar, Phys. Rev. Lett. **92**, 037205 (2004).

[53] G. P. Berman, B. M. Chernobrod, V. N. Gorshkov and V. I. Tsifrinovich, Phys. Rev. B **71**, 184409 (2005).

[54] G. P. Berman, B. M. Chernobrod, V. N. Gorshkov and V. I. Tsifrinovich, cond-mat/0503107.

[55] B. E. Vugmeister, Phys. Stat. Sol. (b) **90**, 711 (1978).

[56] J. A. Sidles, J. L. Garbini, K. J. Bruland, D. Rugar, O. Züger, S. Hoen and C. S. Yannoni, Rev. Mod. Phys. **67**, 249 (1995).

[57] N. Nestle, A. Schaff and W. S. Veeman, Progress in Nuclear Magnetic Resonance Spectroscopy, **38**, 1 (2001).

[58] A. Suter, Progress in Nuclear Magnetic Resonance Spectroscopy, **45**, 239 (2004).

[59] T. E. II Kriewall, *Heterodyne Digital Control and Frequency Estimation in Magnetic Resonance Force Microscopy*, (Dissertation), University of Washington (2005).

[60] T. A. Brun and H. S. Goan, Phys. Rev. A **68**, 032301 (2003).

[61] H. Gassmann, M. S. Choi, H. Yi and C. Bruder, Phys. Rev. B **69**, 115419 (2004).

[62] J. G. Kempf and J. A. Marohn, Phys. Rev. Lett. **90**, 087601 (2003).

[63] S. R. Garner, S. Kuehn, J. M. Dawlaty, N. E. Jenkins, and J. A. Marohn, Appl. Phys. Lett. **84**, 5091 (2004).

[64] R. Budakian, H. J. Mamin, B. W. Chui, and D. Rugar, Science, **307**, 408 (2005).

[65] H. J. Mamin, R. Budakian, B. W. Chui, and D. Rugar, Phys. Rev. B **72**, 024413 (2005).

Index

adiabatic reversal, 2, 12, 13, 14, 65, 66, 85, 89, 90, 91, 100, 106, 129, 160, 161, 164, 207

bandwidth of the measuring device, 36, 38, 63

Bohr magneton, 34, 184

coercivity, 37

collapse time, 92, 94, 160

coherent state, 27, 28, 42, 46, 68, 80, 110, 159, 165

condition of orthogonality, 146

Control-Not gate, 170, 178, 179, 180, 181, 182

CT - environment interaction, 66, 90, 91, 127, 167

damped oscillations, 37, 38, 41, 130

decoherence, 39, 43, 57, 66, 79, 83, 110, 111, 112, 113

decoherence time, 33, 39, 40, 58

density matrix, 42, 43, 44, 46, 49, 79, 80, 81, 110, 111, 112, 113, 207

diffusion coefficient, 40, 93

dipole-dipole interaction, 183, 186, 189, 195, 197

Dirac notation, 26, 27

driven oscillations, 103, 104

E' - centers, 95

effective field, 9, 10, 11, 12, 13, 14, 22, 23, 60, 65, 66, 69, 70, 74, 75, 76, 77, 78, 81, 83, 87, 88, 90, 91, 93, 94, 95, 100, 105, 106, 108, 109, 112, 117, 125, 127, 129, 130, 131, 132, 137, 141, 153, 154, 156, 158, 159, 162, 166, 168, 169

effective mass, 25, 32, 133

eigenfrequencies of the cantilever, 31, 32, 146, 147, 148, 149, 150

eigenfunctions of the cantilever, 31, 32, 146, 148, 149

eigenfunctions of the nonuniform cantilever, 146

electron spin resonance, 1, 10, 184

gyromagnetic ratio, 5, 171, 181

Hamiltonian, 19, 21, 23, 26, 28, 30, 41, 44, 66, 67, 79, 106, 153, 159, 160, 164, 171, 181

Heisenberg
-equations of motion, 24
-representation, 23

interrupted OSCAR technique, 95, 96, 162

Larmor precession, 7, 9, 19, 65

magnetic
energy, 6, 29, 30
noise, 91, 92, 127, 131, 132, 145, 152, 153, 158, 160, 162
susceptibility, 35

magnetization, 35, 131, 134, 137, 138, 174, 183

master equation, 41, 43, 44, 46, 58, 79, 81, 83, 110

ohmic model, 43

operator of
-annihilation, 26, 68
-creation, 26, 27, 68

quality factor, 37, 88, 101, 150

quantum
-collapse, 39, 40, 43, 91, 93, 94, 127, 154, 155, 158, 159, 160, 162, 167
-jump, 64, 83, 91, 92, 93, 94, 95, 96, 127, 129, 130, 145, 153, 154, 155, 156, 157, 158

Rabi
-frequency, 10, 60, 88, 92, 93, 130, 134, 136, 138, 139, 143, 145, 146, 150, 153, 156, 174, 175, 177, 179
-oscillations, 10

resonant slice, 117, 122, 123, 124, 128, 129, 131, 132, 133, 134, 137, 139, 141, 150

rotating system of coordinates, 8, 9, 10, 11, 11, 13, 14, 21, 22, 23, 44, 60, 65, 66, 87, 90, 92, 93, 95, 99, 115, 116, 127, 131, 153, 162, 166, 168

rotational operator, 18

Schrödinger

-cat state, 39, 40, 42, 43, 91, 93, 109, 112, 160, 166, 167, 168

-equation, 19, 22, 28, 66, 67, 160, 164

spin

-diffusion, 3, 183, 188, 190, 192, 193, 196, 203

-flip, 37, 38, 83, 95, 127, 175

- lattice relaxation, 183, 184, 188

-operators, 17, 24, 164

spin - environment interaction, 44, 64, 83, 91

spring constant, 25, 36, 86, 88, 89, 150

tensor product, 46

thermal

-diffusion 42, 53, 66, 79, 81, 83, 110, 111, 112, 113

-vibrations 36, 43, 53, 60, 63, 92, 127, 130, 134, 145, 153

torque, 5, 7

unit matrix, 18

uncertainty relation, 39

Young's modulus, 31

wave function, 20, 22, 23, 28, 39, 40, 42, 43, 52, 73, 74, 94, 107, 108, 109, 154, 155, 159, 160, 162, 164, 166

$\pi-$ pulse, 2, 59, 60, 66, 95, 172, 176, 177, 179, 181